FAO STATISTICAL
DEVELOPMENT
SERIES

14

2000 World Census of Agriculture

Methodological Review (1996-2005)

FOOD AND AGRICULTURE ORGANIZATION OF THE UNITED NATIONS
Rome, 2013

ISBN 978-92-5-107837-2 (print)
E-ISBN 978-92-5-107838-9 (PDF)

TABLE OF CONTENTS

Table of contents

PART TWO
Annex - Detailed tables   75

FOREWORD

Beginning with the 1950 round, FAO has been preparing and advocating decennial Programmes for the World Census of Agriculture (WCA), and the 2000 Programme was the sixth in the series. These Programmes have provided guidance to countries in conducting agricultural censuses following standard concepts, definitions and classifications, and at the same time have promoted the availability of internationally comparable data on the structure of agriculture.

FAO member countries provide the reports of their agricultural censuses to the FAO Statistics Division. Key data on the structure of agriculture and related metadata obtained through the country reports are then disseminated through the FAO website. At the end of each round, FAO publishes a report consolidating the global information about the census conducted during the round. The report contains a compendium of metadata and key data of country censuses, internationally comparable data on selected variables characterizing the structure of agriculture in the world, and a methodological review of the censuses conducted during the round.

For the WCA 2000 round (1996-2005), the traditional contents of WCA reports have been presented in three separate publications. Two of the publications have already been issued. The first publication, a compendium of metadata and key data, was published as FAO Statistical Development Series 12. The second publication in the series, FAO Statistical Development Series 13, presents internationally comparable data on selected variables characterizing the structure of agriculture.

The current publication, which is the third report in the series, relates to a methodological review of the censuses carried out during the WCA 2000 round. Many of the findings included in the report were already used in preparation of the Programme for the WCA 2010 round. Being part of a series of the similar methodological publications since the inception of the WCA Programme, this book serves as a unique reference document for understanding the evolution of strategies and methods of data collection on structural aspects of agriculture. The material presented in this book is expected to serve as research input for further developments in methodologies of agricultural censuses and surveys.

Pietro Gennari

The Director
Statistics Division

ACKNOWLEDGEMENTS

This publication has been possible due to the efforts of a dedicated team of staff and experts led by Mukesh K. Srivastava, Team Leader, Agricultural Censuses and Surveys, who besides conceiving the content and analytical framework of the publication provided technical guidance during the entire process of its preparation.

Basic work of compilation of information for the publication was carried out by Edward Gillin, Senior Statistician, and Paolo Amici, Consultant. Giorgi Kvinikadze, Statistician, produced the first complete draft of this publication, carefully verifying the information from relevant sources, and drawing meaningful conclusions. Adriana Neciu, Consultant, provided research support. The final technical and editorial suggestions made by David Marshall, retired FAO Senior Statistician, led to significant improvement of the report. Tomaso Lezzi and Marianne Sinko carried out the layout of the publication.

ABBREVIATIONS AND ACRONYMS

AH	Agricultural Holding
CAPI	Computer Assisted Personal Interview
CATI	Computer Assisted Telephone Interview
EA	Enumeration Area
EU	European Union
FAO	Food and Agriculture Organization of the United Nations
HH	Household
ICR	Intelligent Character Recognition
IIA	International Institute of Agriculture
OCR	Optical Character Recognition
OMR	Optical Mark Recognition
PC	Population Census
PES	Post Enumeration Survey
PSU	Primary Sampling Unit
SDS	Statistical Development Series
SSU	Secondary Sampling Unit
UNSD	United Nations Statistics Division
WCA	World Census of Agriculture

PART ONE

Report on the 2000 World Census of Agriculture

INTRODUCTION

The first Programme for the World Census of Agriculture (WCA) was developed for the years 1929-1930 by the International Institute of Agriculture (IIA) and implemented in about 60 countries. The 1940 round could not be completed due to the onset of World War II. FAO, having succeeded the IIA, took over the task of organizing the World Census of Agriculture starting from the 1950 round and continued with successive decennial Programmes. The Programme for the WCA 2000 was the eighth decennial international agricultural census Programme.

The Programme for the WCA, on the one hand, serves to promote the availability of internationally comparable data on the structure of agriculture whilst, on the other hand, it provides much needed guidance to countries in collecting the data following standard concepts, definitions and classifications. In many developing countries an agriculture census is a unique source of information on the productive structure of the agricultural sector. An agriculture census lays down the foundations of the agriculture statistics system for a country in the sense that it provides: (1) sampling frames for current surveys and 'ad hoc' specialized surveys, and (2) benchmark data to verify the reliability of data produced by other sources.

With the FAO leadership of the WCA, country participation in this Programme has been increasing over time — from 81 countries in the1950 round to 111 countries in the 1970 round and 122 countries in the 2000 round. Also an increasing degree of regular periodicity of the censuses has been observed among countries, particularly in Europe and Asia. African countries, which earlier depended largely on donor support for the conduct of an agriculture census, are now conscious of the need to conduct an agricultural census at least once every ten years, and are increasingly using their own budget to finance the census.

Out of the 122 countries which conducted an agricultural census during the WCA 2000 round (covering the period 1996–2005), 114 countries provided their reports to FAO. The reports of agricultural censuses received by FAO from member countries serve as the basis for the preparation of internationally comparable data and for studies on methodological issues related to taking an agricultural census.

At the end of each round FAO publishes a report consolidating the global information on structure of agriculture. The report contains a compendium of abstracts of country censuses, internationally comparable data on selected variables characterizing the structure of agriculture in the world, and a methodological review of the censuses conducted during the round.

Keeping in mind the coverage of countries and the available data in the 2000 round, the traditional contents of the report on the WCA 2000 round are now presented in three publications. Two of them have already been issued. The individual country results and the metadata on the country censuses are presented in the publication '2000 World Census of Agriculture: Main Results and Metadata by Country (1996 - 2005)' (FAO, 2010), while internationally comparable data on selected variables are presented in the publication '2000 World Census of Agriculture: Analysis and International Comparisons of Results (1996 - 2005)' (FAO, 2013). The present publication is the third in the series and presents a methodological review of the censuses conducted during the period covered by the WCA 2000 round.

The current publication consists of ten chapters.

Chapter one refers to the basic characteristics of the WCA 2000 Programme. It sets it in the perspective of the previous Programmes and highlights the main changes in methodological developments over time. A summary of the changes in the essential items of the WCA 2000 Programme with respect to the 1990 Programme is provided in Annex Table A1.

Chapter two is centered on the participation of countries in the WCA rounds since the 1930 round. It provides summary information about the number of countries participating in the different rounds. It examines the regularity of census taking and the time gaps between the censuses.

Chapter three investigates different methods and techniques used by countries in census taking. It analyses three main enumeration methods used by countries: complete enumeration, sample enumeration and a combination of both. Various sampling techniques used by countries are summarized to identify the broad trends. Country practices in the use of various techniques for capturing census data (personal interviews, mail, and a combination of both) are also analyzed.

Chapter four reviews the census frames used by countries. It analyses the main types of frames used for complete enumeration and sample enumeration as well as the complete enumeration component of a census conducted by a combination of both methods. Various country practices on the construction of census frames are presented as examples.

Chapter five looks at the extent of the coverage of the different censuses. Ideally a census should cover all holdings in the whole country. However, for different reasons, countries usually restrict the coverage of the census, either excluding some areas (zones/regions) of the country or some types of holdings. Both types of exclusion are analyzed separately.

Chapter six focuses on the census scope and reviews country practices in adapting FAO recommendations to their national situations. Each of the ten categories of census items recommended by the FAO Programme is considered in a separate section.

Chapter seven examines country practices vis-à-vis FAO recommendations with respect to the main classification criteria and the cross-tabulations in census reports. A review of the main data entry and processing software used is also presented.

Chapter eight refers to quality issues related to the census data. Quality checks are classified broadly as: a) quality checks during the preparatory work: b) quality checks during the field work: c) quality checks after the census enumeration. The chapter analyses country practices in applying these types of quality checks.

Chapter nine refers to the dissemination of the census results. It explores the time gaps between the census enumeration and the data disseminations. The use of the web and CD-ROMs to disseminate census results in different ways is specifically analyzed.

Chapter ten analyses the integration of the census of agriculture, during the WCA 2000 round, with other surveys/censuses. Each section is dedicated to country experiences in coordinating the agricultural censuses of the WCA 2000 round with other sources used for the collection of primary data, in particular: population censuses, forestry censuses, aquaculture and/or fishery censuses, and community surveys.

Throughout the whole publication, boxes illustrate particular country examples. A number of summary tables are presented in the text while detailed country level tables are provided in the Annex for reference. The data/information presented in the tables of the Annex have been prepared based on available information. Assumptions have been made only in exceptional cases, and these assumptions have been explained in notes and footnotes, wherever possible.

CHAPTER 1
The Programme for the World Census of Agriculture 2000

1.1 Historical evolution of the Programme for the World Census of Agriculture

While the history of the agricultural censuses starts in 1840 in the United States of America, the idea of a World Census of Agriculture (WCA), a census of agriculture conducted in all countries practically simultaneously, along similar lines and on the basis of the same principles, was conceived only in the early twentieth century by the International Institute of Agriculture (IIA), Rome. The development of this idea led to the first World Census of Agriculture carried out in 1930 under the auspices of the IIA.

It was the intention of the IIA that the WCA 1930 should be the first of a series of World Censuses of Agriculture to be undertaken every ten years. Another WCA, therefore, was planned for 1940. Preparations were made, but the Second World War interfered with the full implementation of the WCA 1940 Programme, leaving it incomplete.

FAO, having succeeded the IIA after the Second World War, launched the WCA 1950 and successive decennial Programmes. Since then, the World Census of Agriculture has become an established international activity. The Programme for the WCA 2000 was the eighth Programme in the series, covering the agricultural censuses to be conducted during the period 1996-2005.

The Programme for the WCA 1930 was designed as the 'Standard Form of Schedule' to be used by all countries. This was a list of items that were considered important for inclusion in a census of agriculture together with some explanatory information. All countries were urged to include in their census all the questions proposed in the Standard Form. The Programme fixed the census year as 1929 for the Northern Hemisphere countries and 1930 for the Southern Hemisphere countries.

Each subsequent Programme was enriched with the experience of the previous Programmes from both the methodological and operational points of view. The next paragraphs of this section briefly describe the evolution of the WCA Programmes from 1940 to 1990.

The WCA 1940 Programme was designed basically in the same format as the 1930 Programme as the 'Standard Form of Schedule'. However, it contained not only a modified form with introductory and explanatory notes, but also recommendations as to the manner of preparing and organizing the census and tabulating and publishing the results. The census year was still fixed – 1939 for the Northern Hemisphere and 1940 for the Southern Hemisphere. This restriction, however, was slightly relaxed, compared to the WCA 1930, allowing countries to choose a year as close as possible to the recommended year in the case of unavoidable circumstances that prohibited the taking of the census in the designated year.

In the WCA 1950 Programme, which was the first Programme developed by FAO, a considerable evolution can be observed regarding the purpose of the census. While the 1930 and 1940 Programmes placed, at the forefront, the determination of accurate and comparable information on the areas under crops, numbers of livestock and quantities of various products entering into the consumption chain, the 1950 Programme brought forward the idea of collecting data on the structural characteristics of agriculture as the primary purpose of the census. This focus on the structural characteristics of agriculture, that change relatively slowly over time, is maintained in all subsequent Programmes. In the 1950 Programme emphasis was put on methodological and operational aspects to make them fit local circumstances. Another new element in the Programme was the existence of both the short list of essential census information, which should be collected by all countries, and an extended list which contained items considered of secondary importance. The 1950 Programme also gave increased attention to the definitions of census items and the tabulation of results.

The WCA 1960 Programme accomplished a quantum jump in census methodology as it introduced the use of sampling methods in census-taking, including post-enumeration sample surveys. The option to use sampling methods increased the number of countries participating in the census of agriculture. The Programme did not, however, introduce substantial changes in the list of items to be collected or their definitions, but it considerably altered the way they were presented. All the proposed census items were arranged into ten sections according to the subject matter and each section consisted of an introduction, proposed items in the short and expanded lists (short list printed in bold type), definitions and explanatory notes, and the tabulation plan. The issue of the relationship between the agricultural and the population censuses was also raised in the 1960 Programme for the first time.

The WCA 1970 Programme introduced the concept of the place of the census of agriculture in the overall system of agricultural statistics, indicating that: (a) agricultural censuses offer an excellent base and framework for planning surveys to produce agricultural statistics; (b) their data can also be used as a bench-mark and as supplementary information for improving the provision of current agricultural statistics; (c) technical and organizational training of the personnel required for carrying out the agricultural census can be exploited for the purpose of organizing other agricultural surveys. The Programme continued to provide guidance on methodological issues related to census-taking. It elaborated on the use of sampling methods at various stages of the census, including their use in pilot censuses and pre-testing surveys, in post-enumeration surveys for checking the quality of data, in controlling processing errors, in tabulation of results, etc. As regards the items to be collected in the census, the 1970 Programme, while retaining the structure of its predecessor, introduced some additional features. Among the major additions was an entirely new section dealing with the association of agricultural holdings with other industries.

The new recommendations of the WCA 1980 Programme were as follows: the census of agriculture should be used as the basis for the collection of current agricultural data through improved methods; it should be utilized for the development and improvement of an overall Programme of food and agricultural statistics; the concepts, definitions and methods should be harmonized with other related statistical systems to ensure comparability and compatibility; well trained and experienced data-collection staff should be retained; more elaborate tabulations, made possible by advances in electronic data processing, should be produced; each country should have greater flexibility and more freedom in adapting the Programme. For the first time, a special chapter was dedicated to guidelines on the preparation and organization of a census of agriculture. The 1980 Programme was the first to explicitly indicate that it referred to national censuses undertaken by countries during the whole decade centred on the reference year of the round.

The WCA 1990 Programme, while recognizing differences in the level of economic and statistical development, encouraged countries to develop and implement the agricultural census according to their national capabilities and requirements. It had three basic characteristics: (a) care was taken that concepts, definitions and classifications used in the census were harmonized with those used in other data sources, both agricultural and non-agricultural; (b) the scope of the data to be collected during the agricultural census was more limited than in previous Programmes; and (c) extensive cross-tabulations were recommended to ensure the maximum use of the census data. The Programme encouraged a complementary relationship between the agricultural census and more frequent food and agricultural sample surveys as well as with related non-agricultural censuses and surveys.

The next section briefly overviews the basic features of the WCA 2000 Programme and the new elements it introduced.

1.2 Basic characteristics of the WCA 2000 Programme

Changes in the WCA 2000 Programme, compared to previous WCA Programmes, were kept to a minimum. The Programme defined a census of agriculture as 'a large-scale periodic statistical operation for the collection of quantitative information on the structure of agriculture'.

The Programme defined three basic objectives of an agricultural census, to:

- Provide aggregate totals for fundamental structural variables that change relatively slowly, to be used as benchmarks for inter-censal estimates;

- Provide a frame for agricultural sample surveys in the form of lists of holdings with some characteristics useful for stratification;
- Provide data for small administrative units and detailed cross-classifications of farm structural attributes.

The WCA 2000 Programme noted that these basic objectives might be too restrictive for some countries. However, limiting the basic objectives in this way came as a direct result of the negative experiences of many developing countries attempting agricultural censuses with wide objectives but without sufficient statistical capacity and resources. For countries with developed statistical systems and sufficient capacity and resources, some additional objectives were, therefore, proposed. The broader objectives included: obtaining benchmark data for improving current crop production estimates, obtaining detailed data on livestock characteristics (such as type, age, sex, breed, use, etc.), and obtaining detailed data on the characteristics of the agricultural population and on the various inputs used for agricultural production. Countries were cautioned, however, that when adding additional census objectives, they should ensure that the basic objectives would not be adversely affected.

The WCA 2000 Programme considered in detail the organizational aspects of census-taking, giving recommendations concerning the establishment of a legal basis for the census, the creation of a census committee, the designation of an agency responsible for conducting the census, budget preparation, publicity, staff recruitment, cartographic preparation, preparation of the holding list, designing the questionnaires and tabulation plan, preparation of field instruction manuals, training of enumerators and supervisors, pre-testing census instruments and conducting a pilot census, the enumeration process, data processing, quality checks and post-enumeration surveys, and the dissemination of the census results.

From the methodological point of view, the WCA 2000 Programme basically followed the recommendations of its predecessors. The agricultural holding was recommended as the statistical unit of the census. It was defined as an economic unit of agricultural production under single management comprising all livestock kept and all land used wholly or partly for agricultural production purposes, without regard to title, legal form or size. The Programme noted, however, that under this approach all land not operated by such agricultural holdings (such as communal grazing land, fallow land under shifting cultivation etc.), which may be important in many countries, would not be covered by the agricultural census.

The activities recommended to be covered by the agricultural census are included in the International Standard Industrial Classification of All Economic Activities, third edition (ISIC) groups 011 (Growing of crops, market gardening, and horticulture), 012 (Farming of animals) and 013 (Growing of crops combined with farming of animals). The WCA 2000 Programme was the first one to provide an explicit link to an international classification in its endeavour to ensure uniformity in concepts, definitions and classifications with those from other data sources. Economic units engaged solely in (a) hunting, trapping and game propagation; (b) forestry and logging; (c) fishing; or (d) agricultural services, were not considered as agricultural holdings and, therefore, considered outside the scope of the census.

The WCA 2000 Programme provided an evaluation of the advantages and disadvantages of using complete or sample enumeration for conducting an agricultural census. It argued that although complete enumeration was desirable if the full objectives of an agricultural census were to be achieved, countries facing capacity and resource constraints should also consider conducting their census on a sample basis. If properly conducted, a sample enumeration may produce even better quality data than a complete enumeration, at the national and higher administrative levels, owing to the reduction of non-sampling errors, though it was also noted that the third objective of providing data for small administrative units could not be easily achieved through sampling. The WCA 2000 Programme, therefore, noted that best solution may be achieved by combining both complete and sample enumeration methods.

Special attention was given to such important issues as census frames, coverage of holdings, data tabulation, and dissemination of census results. The detailed recommendations of the WCA 2000 Programme on these and other topics are considered in the subsequent chapters of the present review.

For the first time the WCA 2000 Programme explicitly introduced the issue of the role of women in agriculture. It admitted that although it was widely recognized that women's participation in agriculture was of great importance, their contribution to agricultural development was in most cases inaccurately reported and often under-estimated. The WCA 2000 Programme, therefore, placed emphasis on the need to present the census results disaggregated by sex, and to take this requirement into account throughout the process of census planning, questionnaire design, data collection, processing and dissemination. It also recommended that even the smallest agricultural holdings should be included in the coverage of the census to ensure that the role of women is properly reflected.

As in previous Programmes, of the whole range of the data items recommended for the census, a subset called 'essential items' was singled out, identified by an asterisk '*'. These items were considered of priority importance for national and international purposes. Definitions and explanations of concepts were generally the same as those in earlier Programmes. Most definitions related exclusively to agricultural statistics. Other general definitions were those recommended by other United Nations agencies to ensure comparability between agricultural census results and data collected through other sources.

There were some changes between the 1990 and 2000 Programmes as regards the list of data items, both essential and recommended. For instance, a new item 'Purpose of production' was introduced in the WCA 2000 Programme as an essential item. Some items concerning irrigation and drainage, recommended to be collected for the whole holding in the 1990 Programme, were recommended to be collected at the parcel level in the 2000 Programme. Annex Table A1 provides a detailed comparative picture of the items included in the 1990 and 2000 WCA Programmes.

An important innovation of the WCA 2000 Programme was the introduction of a set of items, called 'items with environmental implications' and marked by '#', which were supposed to be useful to countries in dealing with environmental issues. This was done in response to a global public and governmental concern about the environment which had been expressed at many international forums, such as 'The United Nations Conference on Environment and Development (UNCED)' and its follow up (Agenda 21). The Programme, however, noted that many of these items were difficult to collect during the census enumeration and may require special qualified training and instruments and, perhaps, specific surveys.

Shortly after the publication of the WCA 2000 Programme, two supplements were issued: 'Guidelines on Employment' (FAO, 1997a) and 'Guidelines on the Collection of Structural Aquaculture Statistics' (FAO, 1997b). The former was a result of the joint efforts of the FAO Statistics Division and the International Labour Office (ILO), Bureau of Statistics. Its preparation was brought about by the necessity to improve the concepts, definitions and standards used to collect employment information in agricultural censuses and make them compatible with recommended international standards. The latter was a recognition that aquaculture (fish farming) had many similarities to crop and livestock farming and was often conducted at the household level. It was intended to assist countries, particularly in Asia, to improve their current surveys in aquaculture and to provide a framework for those countries planning to develop a database on aquaculture. The publication of this supplement was a result of the joint efforts of the FAO Statistics Division and the FAO Fishery Information, Data and Statistics Unit in response to a decision of the Fifteenth Session of the Asia and Pacific Commission on Agricultural Statistics (APCAS) recommending FAO to give consideration to including aquaculture in the Agricultural Census Programme.

CHAPTER 2
Participation of countries in the Programme for the WCA 2000 and previous census rounds

2.1 Historical trends and regularity of census-taking

Many countries have participated in the World Census of Agriculture starting from the 1930 round. Some countries have participated in all complete rounds (1940 is not considered a complete census round), while there are also countries which conducted their first agricultural census during the WCA 2000 round. Some countries which participated in the WCA 2000 round did not exist in earlier census rounds, while other countries ceased to exist as political entities. Also, some territories which are not independent countries, are recorded as separate entities in census reports for statistical purposes[1]. All this complicates comparisons over time.

Annex Table A2 provides the overall picture of agricultural censuses conducted during all complete WCA rounds since 1930. The table presents: 188 FAO member countries as of end of 2005, the end of the WCA 2000 round; Andorra, Brunei Darussalam, Montenegro, the Russian Federation, Singapore and South Sudan, which have become members of FAO since 2005; 14 territories (American Samoa, Bermuda, French Guiana, French Polynesia, Guadeloupe, Guam, Martinique, Montserrat, New Caledonia, Northern Mariana Islands, Puerto Rico, Réunion, Saint Helena, and the US Virgin Islands) which are recorded as separate entities for statistical purposes; some former countries and territories (Alaska, Arab Republic of Yemen, Czechoslovakia, French West Africa, Hawaii, North Borneo, People's Democratic Republic of Yemen, Ryukyu Islands, Saar, Sarawak, Zanzibar/ Pemba, Yugoslavia) which have conducted an independent census during the 2000 round or one of the previous rounds of the WCA. The latter are included in order to be able to compare numbers of participating countries across rounds. For each country, the table shows the participation in all complete rounds of the WCA up to 2000. The year of participation refers to the year in which the census enumeration took place. If, because of several census stages or other reasons, the enumeration took place over several years, all the years are shown separated by a hyphen. If, during a certain round, a country had conducted more than one census, the years of all censuses are shown separated by '&'. For some European countries taking agricultural censuses annually, the reference year of the round is indicated with the word 'annual' in square brackets next to it. During the 1930 and 1950 rounds of the WCA the participation criterion was applied rather flexibly, and those countries (mostly the colonial territories) that could not take a proper census but had made other efforts to obtain the required minimum information were also considered as participants. Such censuses are considered as conducted by special estimates and for them the participation years are marked with '(*)' in Annex Table A2. Variations in nomenclature of countries from one round to another are indicated in footnotes to Annex Table A2.

The grouping of countries by regions follows the standards recommended by the United Nations Statistics Division (UNSD) on the composition of macro-geographical (continental) regions viz. Africa, Americas, Asia, Europe, and Oceania. However, to be able to draw meaningful conclusions from the analysis of country practices, the countries in the Americas have been classified into two groups: 'America, North and Central' and 'America, South' which is consistent with division into sub-regions proposed by the UNSD. This grouping is in conformity with the one used in the earlier publications of the FAO Statistical Development Series.

Table 2.1 below is derived from Annex Table A2 and shows the overall and regional trends in the participation of countries in the WCA rounds. For comparability purposes, censuses conducted by special estimates are not taken into account in the calculation of regional and overall totals.

---

1 Throughout the present publication, for reader's convenience, under a country it will be meant either a country in the political sense of the word, or a territory which for statistical purposes has been considered as a separate entity having participated in a WCA round.

Table 2.1 Number of countries having participated in the WCA rounds by region and by round

Region	WCA round						
	1930	1950	1960	1970	1980	1990	2000
Total number of countries having participated in each round	53	81	100	111	103	94	122
Africa	8	18	29	25	21	23	25
America, North and Central	10	18	18	23	19	16	14
America, South	4	8	11	10	9	7	8
Asia	4	11	20	20	21	14	31
Europe	23	20	17	24	22	21	33
Oceania	4	6	5	9	11	13	11

The table clearly shows that a peak in the number of countries undertaking a census of agriculture was reached during the WCA 2000 round covering the period 1996-2005 with 122 participants, surpassing the last peak of 112 reached in the 1970 round (the period 1966-1975). Europe and Asia showed the biggest increases in the 2000 round relative to previous rounds. In the case of Europe, this was mainly caused by the emergence of new countries after the break-up of Czechoslovakia, the USSR and Yugoslavia in the early 1990s, eight of which either took their first census during the WCA 2000 round or resumed census-taking after a long break. In Asia, three newly emerged countries (Azerbaijan, Georgia and Kyrgyzstan) took their census for the first time but six other countries also undertook an agricultural census for the first time (Bhutan, China, Mongolia, Qatar, United Arab Emirates and Yemen). Afghanistan, Lao People's Democratic Republic and Lebanon also resumed their agricultural census Programme after a long break.

In the other regions, trends in the numbers of countries participating in the WCA are less visible. In Oceania, the upward trend exhibited between 1930 and 1990 appears to have levelled off. In Africa and South America the trend has remained flat for a number of rounds. In the North and Central America Region there is some evidence of a downward trend in the number of countries participating in the census rounds since the 1970 round. One explanation may be that in this region the economic liberalization policies of the 1980s, 1990s and 2000s were quite vigorously applied. As a consequence, a decrease in expenditure and investment in the public sector negatively affected several official activities, in particular public statistical activities. The example of Mexico which has conducted agricultural censuses in all the rounds since 1930, but failed to do so in the WCA 2000 round, supports this hypothesis.

The data of Annex Table A2 enables one to analyse how regularly countries have taken their agricultural censuses. The following simple criteria have been adopted for grouping countries according to their regularity of census-taking. A country is said to conduct agricultural censuses:

a) **regularly** if it participated in **six or seven** out of the seven complete WCA rounds;
b) **with some regularity** if it participated in **four or five** rounds, and
c) **irregularly** if it participated in only **one, two or three** rounds.

There are also countries which have never conducted an agricultural census during the whole period under consideration. Censuses conducted by special estimates are taken into account when analysing the regularity of census-taking.

Annex Table A2 refers to censuses rather than countries. Thus, in view of changes in political boundaries of countries, some censuses in the table refer to several present-day countries. For this reason, in classifying countries according to the regularity of census-taking, the following criterion has been adopted: *if the whole territory of a present-day country was covered during a round as a part of another country or countries, it is deemed that the country has participated in that round.* All relevant information is in the footnotes to the table. According to this approach, the census of 1929-30 of India is attributed to Bangladesh, India and Pakistan. Likewise, all censuses conducted by the former Czechoslovakia and Yugoslavia are attributed to all the countries that emerged after their break-up, the 1929-30 census in French West Africa (conducted by special estimates) is attributed to all of the countries forming that territory at that period, and so on.

Table 2.2 below summarizes the regularity of census-taking according to the above criteria.

Table 2.2 Numbers of countries by regularity in census-taking

Region	Total number of countries	Regularly	With some regularity	Irregularly	Never
WORLD #	208	48	55	84	21
Africa	55	5	15	33	3
America, North and Central	29	9	13	7	0
America, South	13	3	6	4	0
Asia	47	6	9	22	10
Europe	42	21	8	6	7
Oceania	21	4	4	12	1

\# World means 'all FAO member countries as at end 2005 plus Andorra, Brunei Darussalam, Montenegro, the Russian Federation, Singapore and South Sudan plus other territories that have ever conducted an independent agricultural census'

The table shows that, during the period under review, about half the countries in the world had either undertaken their agricultural census on an irregular basis (84 countries) or had never conducted an agricultural census (21 countries). Both the African and Asian regions exceed this average share with a high proportion of countries taking their agricultural census on an irregular basis. In the Asia region there are still ten countries that have never taken an agricultural census and the number of those only taking a census on an irregular basis is also high. At the same time, there are countries which have conducted agricultural censuses regularly: India, Indonesia, Japan, Pakistan, Philippines, Sri Lanka, and Turkey, with India and Japan having established a five-yearly cycle of agricultural census Programmes.

In the Americas (North, Central and South), about two thirds of the countries conducted agricultural censuses either in all census rounds or with a high degree of regularity. Europe, however, was the leader in the regularity of census-taking, with half the countries conducting agricultural censuses on a regular basis (in all rounds). This is largely due to EU regulations establishing a regular Programme of agricultural censuses and surveys for collecting agricultural statistics to serve the needs of the Common Agricultural Policy, among other things.

2.2. Participation of the countries in the WCA 2000 round

Altogether, 122 countries participated in the 2000 round as shown in the Map 1.

It was recommended in the WCA 2000 Programme that countries should take their agricultural census as close to the year 2000 as possible. To assess how strictly countries followed this recommendation, the census round (1996-2005) was divided into three parts:

a) **beginning** of the round – 1996-1998;
b) **middle** of the round – 1999-2002;
c) **end** of the round – 2003-2005.

Map 1.
Countries that conducted an agricultural census during the WCA 2000 round (1996-2005)

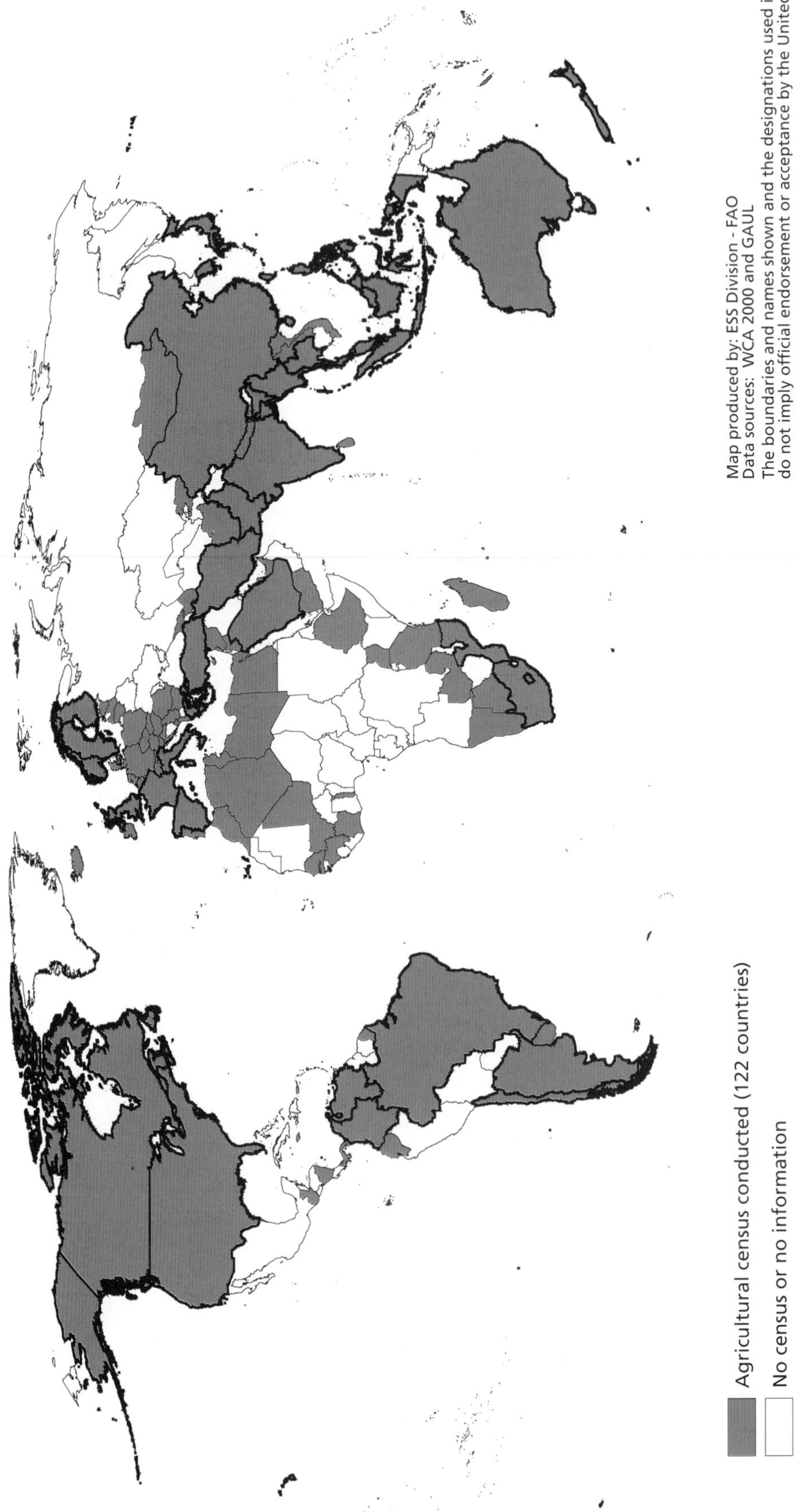

Map produced by: ESS Division - FAO
Data sources: WCA 2000 and GAUL
The boundaries and names shown and the designations used in this map
do not imply official endorsement or acceptance by the United Nations.

Agricultural census conducted (122 countries)

No census or no information

Table 2.3 below shows the distribution of participating countries by their period of participation within the round.

Table 2.3 Timing of censuses during the WCA 2000 round

Region	Total number of countries	Number of countries that conducted the census in the		
		Beginning (1996-1998)	Middle (1999-2002)	End (2003-2005)
All participating countries	122	12	79	31
Africa	25	3	15	7
America, North and Central	14	2	10	2
America, South	8	3	5	0
Asia	31	3	13	15
Europe	33	1	28	4
Oceania	11	0	8	3

As seen from the table, about two thirds of participating countries conducted their census in the middle of the decade covered by the WCA 2000 round, thus following the recommendations of the Programme. Of those countries which did not conduct a census in the middle of the period, more than two thirds conducted a census at the end of the round. This is especially true for Asia, where more countries conducted a census at the end of the round than in the middle.

The case of Europe should be especially mentioned where practically all participating countries conducted their censuses close to the reference year of the round. To a considerable extent this is due to EU regulations obliging member countries to conduct an agricultural census in the years ending with nine or zero. The only exclusions were the non-EU countries Albania, Croatia and Montenegro, as well as Bulgaria and Lithuania which became EU members after they conducted their agricultural censuses.

Of the 122 countries that conducted an agricultural census in the WCA 2000 round, only 114 have presented a report to FAO. The analysis that follows in this review relates to those 114 countries.

Some time-related information is associated with an agricultural census, such as year(s) of participation, time gap from the previous census, reference period, reference day and year of publication of the census results. Annex Table A3 presents these characteristics for the reporting countries. In the calculation of time gaps, some simplifying assumptions were made, specifically: (a) if a country has conducted several censuses during the WCA 2000 round, the time gap was calculated between the last two censuses of the round; (b) if a census was taken during more than one year, the last year of the census (when it was actually completed) was considered in the calculation of the time gap; (c) censuses conducted by special estimates were ignored.

Examination of Annex Table A3 reveals that eight countries out of 114 conducted the first census of agriculture on their territory during the WCA 2000 round. Three of them (Azerbaijan, Georgia, Kyrgyzstan) were newly emerged states following the break-up of the USSR which had never participated in the WCA Programme. Others were the Comoros and Gambia in Africa, and Bhutan, China, and Qatar in Asia. Mongolia has been conducting annual livestock censuses since 1924 but the census of 2000 reported in Annex Table A3 is the first one that has been included in an FAO review. Thus it is marked as 'First reported census'. Croatia and Serbia, as well as the Czech Republic and Slovakia conducted their first agricultural censuses as independent countries but an agricultural census was previously conducted on their territories when they formed a part of another country (Yugoslavia and Czechoslovakia, respectively). Thus they were not considered as countries having conducted a first census and for them the time gap was calculated between the census conducted in the WCA 2000 round and the last census of the country of which they were a part. Likewise, Yemen conducted its first agricultural census as a united country in 2003 but its whole territory was covered in the 1980 round by the censuses conducted in Yemen Arab Republic and Yemen People's Democratic Republic, in 1983 and 1984, respectively, so the time gap for Yemen was taken as 19 years.

Of the participating but not reporting countries, mention should be made of: a) Montenegro which conducted its first agricultural census in 2003 as an autonomous unit within the state of Serbia and Montenegro but its territory was covered by the 1981 agricultural census of Yugoslavia; b) Kiribati and Tuvalu which conducted their first agricultural censuses (jointly with the population census), not counting special estimates in the 1930 and 1950 rounds; and c) United Arab Emirates which also conducted the first agricultural census in their history in 2004.

Table 2.4 below is derived from Annex Table A3. It shows for each region the typical (modal, i.e. most frequent) and median time gaps from the previous census of agriculture, as well as the extent of variation in the time gap between censuses. The number in parentheses after the mode shows its frequency. Countries that conducted their first census were not taken into account when compiling the table.

Table 2.4 Mode, median and range of variation of the time gap between the last two censuses

Region	Time gap (years)			
	Mode (frequency)	Median	Max	Min
All reporting countries	10 (19)	11	73	1
Africa	9,10,11 (3)	11	50	2
America, North and Central	5 (4)	12	30	5
America, South	14,16 (2)	14	19	10
Asia	10 (6)	10	36	5
Europe	10 (6)	10	73	1
Oceania	5 (3)	10	16	5

As seen from the table, countries typically follow the ten-year cycle. The unusually high maximum figure for Europe is due to the fact that Estonia, Latvia and Lithuania conducted their first censuses since the 1930 round after regaining independence in the early 1990s.

CHAPTER 3
Enumeration methods and data collection techniques

This chapter consists of three sections. Section 3.1 examines country practices on the methods of census enumeration, namely: complete enumeration, sample enumeration and a combination of both. Section 3.2 reviews the main sampling designs used for censuses conducted using sample enumeration techniques. Section 3.3 looks at the different techniques of data collection used by countries, like direct interview, postal enquiries, and a combination of both; the use of objective measurements is also reviewed.

3.1 Enumeration methods: complete versus sample enumeration

As noted in the WCA 2000 Programme, 'The word 'census' implies a complete enumeration of all agricultural holdings. However, by extension, it can be conducted by a sample enumeration, provided the sample is large enough to generate sub-national data'. Many countries have chosen the option of sample enumeration, either exclusively, or in combination with complete enumeration. Usually budget constraints were cited as the main reason behind such a choice. A detailed discussion of the advantages and disadvantages of complete and sample enumerations can be found in FAO, 1996a and FAO, 2005. The present section analyses country practices in this respect that were observed in the WCA 2000 round.

Annex Table A4 provides detailed information for each of the 114 reporting countries on the enumeration method used, as well as sampling designs in cases where sampling was used either exclusively or in combination with complete enumeration. Footnotes clarify all cases of combined complete and sample enumerations (such cases are marked by '√' in both 'Complete enumeration' and 'Sample enumeration' columns). Table 3.1 summarizes information on the enumeration methods used.

Table 3.1 Number of censuses of the WCA 2000 round by enumeration method

Region	Total	Enumeration Method		
		Complete enumeration (a)	Sample enumeration (b)	Combination of (a) and (b)
All reporting countries	114	76	16	22
Africa	25	8	10	7
America, North and Central	14	12	0	2
America, South	8	6	0	2
Asia	29	16	5	8
Europe	29	27	0	2
Oceania	9	7	1	1

Two thirds of the reported agricultural censuses of the WCA 2000 round were carried out by complete enumeration. Complete enumeration censuses were quite common in Europe, North and Central America, Oceania, and South America where over 75 percent of reporting countries followed this practice.

Out of the 16 countries that conducted the census on a purely sample basis, ten were in Africa. In six out of seven censuses in Africa conducted by a combination of complete enumeration and sampling, complete enumeration covered only the large holdings. The bulk of holdings in these censuses were enumerated on a sample basis, and in fact these censuses may be considered as sample enumeration censuses with one complete enumeration stratum. Of the eight African countries which conducted a complete enumeration census, three (Algeria, Egypt and Libya) were from the more developed North Africa, two (Cape Verde and Reunion) are small islands and three (Seychelles, Uganda and Zambia) conducted their agricultural census as a module of their population and housing census. This scenario leads to a conclusion that sample enumeration was the most common method for African countries to conduct an agricultural census in the WCA 2000 round.

The methods of combining complete and sample enumeration practiced by countries can be grouped into three main categories as follows:

1. The first type of combination refers to cases where large or important special holdings are enumerated by complete enumeration while the small and medium sized holdings are covered by sample enumeration. Such cases, in a sense, may be treated as special cases of sample enumeration with one stratum sampled with 100 percent coverage. Nearly half of the 21 censuses adopting a combination of sampling and complete enumeration fall in this category. They include six African countries (Botswana, Cote d'Ivoire, Mali, Mozambique, Tanzania, and Tunisia), Jamaica in North and Central America, Ecuador and Colombia in South America, and Azerbaijan in Asia. The agricultural census of Turkey, where villages and settlements with 5 000 and more inhabitants were completely enumerated, may also be classified in this category.

2. The second type of combination is closer to complete enumeration and refers to cases where most of agricultural production is covered by complete enumeration but one stratum of holdings (below a certain threshold or considered small in some other sense) are enumerated on a sample basis (maybe with a smaller questionnaire) to ensure a complete picture of agriculture. Censuses of this type were conducted by two European countries: Slovakia and the United Kingdom, and Kyrgyzstan in Asia. Slovakia and the United Kingdom covered holdings below a certain threshold by sample enumeration, while Kyrgyzstan surveyed a 35 percent sample of small household plots owned by dwellers in both urban and rural areas. The case of Slovakia is described in more detail in Box 5.1.

3. The third type refers to cases where a small questionnaire is applied to all holdings while a more detailed one to a sample. In total, eight censuses were reported to have been conducted this way. Enumeration of this type was most popular in Asia where Afghanistan, India, Lao People's Democratic Republic, and Yemen used this method. This type of enumeration was also used by Morocco, Samoa and the United States of America. In India and Yemen a complete enumeration was carried out as part of the listing activities of the agricultural census and some basic data were collected during this operation (land area of the holding and sex and social group of the holder in India; land area and livestock numbers in Yemen). This ensured a complete frame for subsequent sampling while at the same time provided basic agricultural data for tabulation at the lowest administrative level.

The Viet Nam census was also conducted by a combination of complete and sample enumeration but the information on the exact design of the survey was not available at the time of writing this review.

3.2 Sample designs

A total of 38 countries used sample enumeration, either exclusively or in combination with a complete enumeration.

One way of grouping the sample designs is to classify them on the basis of type of sampling frame used, i.e. lists or land area segments. List and area sampling frames are discussed in Section 4.1 of Chapter 4. Annex Table A4 indicates that only two South American countries – Colombia and Ecuador used an area sample design for their censuses in the WCA 2000 round. Both censuses were conducted by combination of complete and sample enumeration, and area frames were used for their sample enumeration components. Some countries use area frames for their current agricultural surveys which are not covered by this review.

Another way of classifying sample designs is by categorising them by the number of stages of sampling used in the design.

In case of the list sample design, the one-stage design refers to the case where the units of interest for the census (holdings to be enumerated or households to be screened for holdings) are selected directly from a list frame, while multiple stage designs refer to the cases where holdings or households are selected at the last stage of the sample selection process after first selecting primary sampling units (PSUs), then selecting secondary sampling units (SSUs) from the selected PSUs, and so on. Usually, holdings or households are SSUs but they may be also tertiary and even fourth order units as in case of Pakistan described in Box 3.1. In the case where all holdings or households within the last selected units are enumerated, the number of stages so defined will differ from the number of actual sampling procedures. For example, in Tunisia a sample of Enumeration Areas (PSUs) was selected and all holdings (SSUs) within selected PSUs were enumerated, so actually one sample selection was carried out. From the sampling theory point of view, this is an example of one-stage cluster sampling. The same is true for India and Philippines where villages ('barangays' in the Philippines) were selected randomly and all holdings/households within the selected villages were visited. However, all three cases are classified as multi-stage sampling in Annex Table A4 because the units selected at the first stage were not the ones enumerated during the census. In such cases the secondary sampling units may be assumed to be selected with probability one. All such cases are indicated in the footnotes to Annex Table A4 by highlighting the words 'cluster sampling'.

In the case of area sample design the ultimate units to be selected are segments of land, with selection probabilities proportional to their area. The segments are then associated with the holdings to be enumerated. Usually the segments are selected through a two-stage procedure where first larger land areas with recognizable boundaries, called frame units, are selected with probability proportional to their area, then the selected areas are divided into equal segments and at the second stage the segments are selected with equal probability.

Table 3.2 below is derived from Annex Table A4 and shows the sampling practices as defined by the number of stages discussed above.

Table 3.2 Number of censuses of the WCA 2000 round by sample design

	Total	List sample design			Area sample design
		Stages			Stages
		One stage	Multiple stage	No info	Multiple stage
All censuses with sample enumeration	37	6	27	3	2
Africa	17	0	16	1	0
America, North and Central	2	2	0	0	0
America, South	2	0	0	0	2
Asia	13	0	11	2	0
Europe	2	2	0	0	0
Oceania	2	2	0	0	0

As seen from Table 3.2, out of 33 countries reporting the use of the list sample design, only six countries used a one-stage sample. This is not surprising since one-stage sampling is appropriate only when an exhaustive up-to date list of holdings exists. Of the six countries, four (New Zealand, Slovakia, the United Kingdom, and the United States of America) have sufficiently developed statistical systems for creating such lists while the remaining two (Jamaica and Samoa) made special efforts to create such lists by enumerating all households during the first phase of the census.

All the 27 countries reporting multiple stage list sample designs were from Africa or Asia. Further sampling by stages allows for greater flexibility to enhance the efficiency of the sampling design. In all cases except Pakistan this was in fact a two-stage design with enumeration areas or villages (or analogous administrative subdivisions) used as PSUs and agricultural holdings or households screened for holdings as SSUs. Out of these 27 cases, PSUs were sampled with probability proportional to size in 18 censuses and with equal probability in five censuses, while for four censuses information about the selection method of PSUs is not available in the census reports.

The SSUs for two-stage sample designs were agricultural holdings or households subsequently screened for holdings during the enumeration. There was a clear preference for equal probability sampling (either simple random or systematic) for selection of SSUs, among the reported cases, of which in three cases all SSUs within the selected PSUs were enumerated.

Box 3.1 Examples of sample design

One-stage sample design: USA 2002

The official Census Mail List was established on 1 September 2002. The list contained 2 841 788 records. Two census forms were elaborated: the non-sample form and the sample form, the latter containing some additional questions.

The sample form was mailed to all mail list records in Alaska and Rhode Island and to a sample of records in other states. Mail list records were selected into the sample with certainty if they (1) were expected to have large total value of agricultural products sold or large acreage, (2) were in a county with less than 100 farms in 1997, or (3) had other special characteristics (e.g., abnormal farms such as institutional farms, experimental and research farms, Indian reservations, etc.).

Mail list records were systematically sampled: for counties containing 100 to 199 farms in 1997 at a rate of one in two, for counties containing 200 to 299 farms in 1997 at a rate of one in four, for counties containing 300 to 399 farms in 1997 at a rate of one in six, and for counties containing 400 or more farms in 1997 at a rate of one in eight. The mail list records not in the sample received the non-sample form.

Two-stage sample design: Senegal 1998-99

The sampling procedure at the first stage was stratified. A department (an administrative unit of Senegal) was considered as one stratum if it was deemed to be homogenous from the agro-ecological point of view. Otherwise it was divided into two or three strata (six departments were divided in total).

At the first stage, enumeration areas (districts de recensement) were selected with replacement and probability proportional to the number of agricultural holdings – 30 in large strata, 25 in medium and small strata and 15 in one small department of Rufisque. At the second stage, in each enumeration area ten agricultural holdings were selected with equal probability. In total 7 250 agricultural holdings were selected which constituted 1.66% of the total population of agricultural holdings.

Three-stage cluster sample design: Pakistan 2000

Pakistan used three stage cluster sample design in rural settled areas of North Western Frontier, Punjab and Sindh Provinces.

At the first stage, clusters of villages (mouzas) called Patwar Circles were selected with probability proportional to size. At the second stage two mouzas were selected with probability proportional to size in each selected Patwar Circle. The sizes of Patwar Circles and mouzas were defined based on the number of households and cultivated area. At the third stage, clusters of about 30 households were selected with equal probabilities in each selected mouza. The ultimate sample included all households from the clusters selected at the third stage of sample selection. Thus the households selected were fourth order units selected with probability one.

3.3 Data collection techniques

Data collection techniques refer to the various ways of gathering the information. During the WCA 2000 round, direct personal interviews, postal enquiries or a combination of both methods were used. Annex Table A5 shows data collection techniques used by countries during the WCA 2000 round. Wherever a combination of direct interview and mail techniques was used, a footnote clarifies the type of the combination. No country reported the use of computer assisted methods (CAPI or CATI) or the internet for data collection.

However, the period covered by this review witnessed a keen interest to try these new technologies. Some countries used them as auxiliary tools, e.g. the United States of America used CATI for collecting data from certain groups of non-respondents.

Table 3.3 below summarizes the different data collection techniques used by countries. It can be seen from the table that direct interview was the most common data collection technique with over 80 percent of the reported censuses undertaken using this method alone. In South America the direct interview method was used in all the eight censuses covered in the present review. In Asia and Africa all but one census were conducted by the direct interview method, in Oceania all but two and in North and Central America all but three. The main constraint for using the cheaper mailing techniques were insufficiently developed postal services, low education level of holders, and imperfectness of farm registers or frames which increases the risk of under-coverage of new holdings which can be identified more easily by enumerators in the field.

Only six countries used purely mail enumeration techniques during the WCA 2000 round: Denmark, Ireland, Norway, Sweden and the United Kingdom in Europe and New Zealand in Oceania. The combination of the two techniques was, however, more popular with 12 countries using it. There were two main types of combination of interviewing and mailing techniques. The first type refers to the cases where the questionnaire was mailed to all holders and interviewing was used as a follow-up for those holders who did not respond to the mailed questionnaire. This method was used by Australia, Canada, Netherlands and Puerto Rico. The second type refers to the case where questionnaires were mailed to those holdings considered more capable to respond by mail (mostly large farms and business entities), with direct interview follow-up if required, while the remaining holdings were surveyed by the direct interview method. This method was used by Botswana, Croatia, Czech Republic, Hungary, Slovenia, Sri Lanka and the United States of America. In Botswana, for instance, the questionnaires were mailed to holdings in the commercial sector; however, because of the high non-response rate, substantial follow-ups were carried out. The techniques used by Malta could be classified into this group but for the following circumstance: in Malta mail questionnaires were sent to holdings having less than 3 ha of dry land and no means of irrigation while all others were enumerated by personal interviews. The reason for choosing the mail approach for this group of holdings was perhaps the greater tolerance of non-response that could be accepted from this category of holdings on account of their expected low impact on the overall census results. An interesting combination of interviewing and mailing techniques was used in Finland where questionnaires were sent by mail and the data were collected by telephone interviews.

Table 3.3 Number of censuses of the WCA 2000 round by different techniques for collecting the census data

Region	Total	Interview (a)	Mail (b)	Combination of (a) and (b)	Use of objective measurement	
					Only Area	Area & Yield
All reporting countries	114	96	6	12	5	8
Africa	25	24	0	1	1	7
America, North and Central	14	11	0	3	0	0
America, South	8	8	0	0	0	1
Asia	29	28	0	1	3	0
Europe	29	17	5	7	1	0
Oceania	9	7	1	1	0	0

In some countries the enumerators carried out the objective measurement of areas and/or yields during the visit to the holding. The objective measurement of areas and yields (crop-cutting) were mostly used when the holder's response was considered unreliable or there were no official records of the land area of the parcels being cultivated. However, this was a very resource-intensive method of data collection and was usually used where the census was conducted on a sample enumeration basis. Only 13 countries actually used objective measurements during census-taking in the WCA 2000 round (eight of them in Africa). In all but two of these cases (Czech Republic and Kyrgyzstan) the census was taken by sampling method.

CHAPTER 4
Census frames

4.1 Area frames and list frames

A census frame provides identification of the statistical units of the population of interest (i.e. agricultural holdings). It should cover all holdings in the country without omission and without duplication and is a pre-requisite for both complete or sample enumeration censuses. The two types of frame used for conducting an agricultural census are the area frame and the list frame. A combination of list and area frames is referred to as a multiple frame.

Area frame is normally used for sample enumeration, and is more commonly used for agricultural surveys than agricultural censuses. It is an ordered list of land areas, called frame units, with an assigned number of approximately equal segments (smaller land areas), which cover the entire territory to be surveyed without overlap or omission. Land use strata are generally included in the survey design with the most intensely cropped areas being sampled more heavily than areas with little or no agriculture. Land units from the area frame are usually selected with probability proportional to their area . The first stage land units are selected according to probability proportional to size and the second stage units (segments) are identified and selected from the selected frame units (PSUs) with equal probability. Maps, satellite imagery and aerial photographs are usually used for construction of an area frame. There are rules developed to associate holdings with each segment, and ultimately the data are collected from the holdings associated with the selected segments.

List frames are formed by compiling lists of holdings including identification particulars of the holder. Often the frame contains auxiliary information on the holdings as well. The auxiliary information, such as land area operated, number of livestock held, by species, is useful for stratification, determination of sample size and its allocation to strata or clusters. For efficient management of the field work, the list frames are arranged by enumeration areas (EAs). An EA is a well-defined geographical territory, usually assigned to one enumerator for the collection of data during the census period. For complete enumeration censuses the EAs are just a convenient tool for organizing the field work, while for the censuses conducted on sample basis they often serve as PSUs.

Area and list frames are discussed in detail in FAO, 1996a and FAO, 1996b.

An ideal list frame would be a complete list of all holdings, i.e. agricultural farms qualifying as an 'agricultural holding' in accordance with the operational definition of the census unit, and should be up-to-date on the census reference date. Some countries maintain and regularly update registers of holdings (farm registers). If the process of tracking the emergence and disappearance of holdings (births and deaths) and updating the registers on an on-going basis is in place and reliable, these registers may be close to an ideal frame. However, the workload, challenges and related costs involved in establishing and regularly updating a farm register are such that these registers have succeeded only in statistically developed countries where supported by an appropriate legal framework. These permanent registers have succeeded usually in cases where a link with some administrative process has been established for updating. Nonetheless, countries need to verify the validity of such registers just before the census or undertake a listing operation to create a specific frame for the census. Some countries use their population census, if carried out shortly before the agricultural census, to create a frame for the agricultural census.

Often it is found useful to use the same enumeration areas (EAs) for the agricultural census as used for the population censuses. In this case a frame of EAs is already available for complete coverage or for drawing a sample. The frame of EAs cannot be considered as an area frame because enumeration areas are not selected with probability proportional to their physical area, but with probability proportional to their population, number of households, number of holdings etc., or even with equal probability.

An area frame is often combined with a list frame of large and/or special holdings. In these cases the area frame is used to collect data on the small and medium sized holdings on a sample basis while the large and/or special holdings are included on a complete enumeration basis. Such a design is called a multiple frame design. In the case of multiple frame designs, special care is needed to remove list frame elements from the area frame to avoid double counting.

4.2 Classification of frames used in the WCA 2000 round

Annex Table A6 describes the types of frames used by the reporting countries in the WCA 2000 round. The country practices illustrate the use of the different types of frames mentioned in the previous section. Only two reporting countries, Columbia and Ecuador, used area frames for census enumeration. Both countries used the multiple frame approach – the area frame was used in combination with a list of special holdings which were enumerated on a complete enumeration basis. Puerto Rico provided an interesting example of using a sample from the area frame for adjusting the results of the complete enumeration of the list frame for coverage errors. This example is discussed in more detail in chapter eight dealing with quality assurance issues.

The list frames used by the reporting countries can be classified into two broad categories:

I.   Frames where a list of agricultural holdings was available before the census enumeration;
II. Frames where the list of agricultural holdings was created during the enumeration process.

The frames from Category I may be further subdivided into four classes:

1. A farm register maintained by the country on a regular basis. As mentioned above, such registers exist in quite a few countries. In Annex Table A6 this type of frame is designated 'Maintained farm register'.
2. A list of holdings compiled using various sources like administrative sources (business registers, land cadastres, tax registers, etc.), statistical sources (statistical registers maintained by statistical agencies, lists from previous agricultural censuses and surveys etc.), and other non-administrative sources. In Annex Table A6 this type of frame is designated 'Administrative sources' if the frame was compiled based exclusively on administrative sources, 'Non-administrative sources' if the frame was compiled based exclusively on non-administrative sources, and 'Adm. & non-adm. sources' if the frame was compiled based on both types of sources.
3. A list of holdings prepared during some listing activities conducted either by the census team as a preliminary phase of the census or by local government officials prior to the census enumeration. In Annex Table A6 this type of frame is designated 'Pre-listing activities'.
4. A list of agricultural holdings created during a population census conducted prior to the agricultural census, where the census questionnaire contained some questions enabling the identification of agricultural holdings. In Annex Table A6 this type of frame is designated 'List of AH from a PC'.

The frames from Category II may be subdivided into two broad classes:

1. Only a list of territorial units (mostly enumeration areas but sometimes also administrative units like villages, communes, and districts) was available from cartographic materials or administrative sources, and the list of holdings was created during the census enumeration by screening these territorial units. In Annex Table A6 this type of frame is designated 'Screening EAs', 'Screening villages', 'Screening communes' or 'Screening districts' depending on the type of territorial units screened. In case of multi-stage sampling the word 'selected' is added.
2. A list of households was available from a population census conducted recently or jointly with the agricultural census, and the list of holdings was created during the census enumeration by screening these households. In Annex Table A6 this type of frame is designated 'List of HH from a PC' if the population census was conducted prior to the agricultural census and 'Population census module' if agricultural and population censuses were conducted as a joint activity.

4.3 Frames used for censuses conducted exclusively by complete enumeration and frames used for sampling components of censuses.

Table 4.1 below summarizes the types of census frames according to the above classifications/classes for censuses conducted exclusively on a complete enumeration basis, while Table 4.2 summarizes the types of sampling frames used in the censuses which used sampling enumeration, either exclusively or in combination with complete enumeration.

Table 4.1 - Number of complete enumeration censuses by type of census frame

Region	Total	Frames available prior to enumeration				Frames created during the enumeration	
		Maintained up-to-date farm register	By pre-listing activities	By compilation from various sources (including administrative)	From a prior population census with screening questions	By screening territorial units (EAs, villages etc.)	By screening a list of households from a population census (prior or simultaneous)
All reporting countries	76	8	11	19	3	27	8
Africa	8	0	3	0	1	1	3
America, North and Central	12	1	1	3	1	5	1
America, South	6	0	1	1	0	4	0
Asia	16	0	4	1	0	8	3
Europe	27	6	2	11	1	6	1
Oceania	7	1	0	3	0	3	0

Table 4.2 Number of sample censuses by type of sampling frame

Region	Total	Frames available prior to enumeration					Frames created during the enumeration		No info
		Maintained up-to-date farm register	By prelisting activities	By compilation from various sources	From a prior population census with screening questions	Area frame	By screening selected territorial units	By screening a list of households from a population census	
All reporting countries	38	2	7	7	3	2	13	1	3
Africa	17	0	2	1	2	0	11	0	1
America, North and Central	2	0	1	1	0	0	0	0	0
America, South	2	0	0	0	0	2	0	0	0
Asia	13	0	3	4	1	0	2	1	2
Europe	2	1	0	1	0	0	0	0	0
Oceania	2	1	1	0	0	0	0	0	0

As seen from those tables, only ten countries out of 114 reporting used a maintained up-to date farm register as a frame, of which seven were in Europe (Austria, Belgium, Denmark, Netherlands, Norway, Sweden and the United Kingdom), two in Oceania (Australia and New Zealand) and one in North America (Canada). Of these countries, eight conducted a complete enumeration census, while the remaining two used the farm register for drawing a one-stage sample of holdings for enumeration (in New Zealand the whole population was sampled while in the United Kingdom just small holdings). A well maintained farm register provides an ideal frame for one-stage sampling, and in combination with mailing techniques of data collection it provides a fast and cost-effective option for conducting a census. Eight out of the above mentioned countries conducted their census using mailing techniques and such was the relatively low cost of this approach that four of them conducted agricultural censuses annually. The feasibility of conducting an annual census also depends to some extent on the coverage, scope and number of holdings to be covered in the census.

The most popular way of obtaining a frame for the census was the creation of a list frame during the enumeration process itself by screening territorial units (all units in case of complete enumeration and selected units in case of sample enumeration). About a third of the countries covered by this review (37 countries) used such frames. The territorial units screened were mostly enumeration areas but sometimes different units were also adopted for screening: all districts in Bhutan; selected villages/settlements in Comoros; all communes in Luxembourg; selected 'fokontanys' in Madagascar; selected villages, wards, zones, town quarters in Mozambique; all 'bags' in Mongolia; selected 'barangays' in Philippines; all villages and agricultural complexes in Saudi Arabia; and selected villages in Turkey.

Six countries used the frame of households from their population census (conducted simultaneously or shortly before) as a starting point for the identification of agricultural holdings. These cases will be considered in more detail in Chapter ten containing a review of the integration of the agricultural census with the population census.

It should be noted that in Table 4.2, all cases in Africa, South America and Asia refer to multiple stage sampling while all cases in North and Central America, Europe and Oceania refer to one-stage sampling. It can be seen that only six countries have used one-stage sampling in their agricultural censuses. This is quite natural because one-stage sampling requires the availability of an exhaustive list of all agricultural holdings before the enumeration operations are started, which is not always feasible. As mentioned above, New Zealand conducted the agricultural census exclusively on a one-stage sample basis, using a farm register as a census frame. The United Kingdom also used a farm register but conducted its agricultural census by combining complete enumeration of larger farms with one-stage sample enumeration of smaller farms. Slovakia and the United States of America also used the approach of combining complete enumeration with one-stage sampling but compiled their census frames from various administrative and non-administrative sources. Finally, Jamaica and Samoa, which also conducted their censuses by a combination of complete and one-stage sample enumeration, created the census frame during the first phase of the agricultural census by visiting all households for the identification of agricultural holdings.

4.4 Frames used for complete enumeration components of the censuses with a sampling component

Analysing the frames used for complete enumeration components of the censuses conducted by combination of complete and sample enumeration, one should take into account the classification of such censuses given in Section 3.1 of Chapter three.

The first type of combination census refers to the case where a small number of large and/or special holdings are enumerated completely and the bulk of the holdings are covered by a sample. For ten out of 11 censuses of this type, the frame used for the complete enumeration component was compiled from administrative and/or non-administrative sources and was different from the frame used for the sample enumeration component. The only exception was in Jamaica where the frame was created during the pre-listing activities and used for both the complete and sample enumeration components.

For the second type of combination, where the sample was used only for the investigation of holdings that were below the established threshold or were considered small in some other sense, the same frame was used for both components (a list compiled from administrative sources in Kyrgyzstan, a list compiled from administrative and non-administrative sources in Slovakia, and a farm register in the United Kingdom).

The third type of combination refers to the case where all holdings are screened using a 'short' questionnaire to collect some basic data and to create the required list frame(s) and a sample is surveyed with a 'long' questionnaire for collecting more detailed data. All these censuses used the same frame for both components. However, the process of frame construction was different. For the censuses of India, Lao Peoples Democratic Republic, Samoa and the United States of America the frame was compiled prior to the enumeration and then used for both components, while for the censuses of Afghanistan and Yemen the frame was created during the complete enumeration phase and then used for the sample enumeration phase. In the cases of Afghanistan and Yemen the same frame falls into two different types according to the adopted classification: the complete enumeration frame was created by screening enumeration areas and it is classified into 'Screening EAs', while when used for the sampling component the same frame is classified into 'Pre-listing activities' because the listing activities were conducted by the census team during the preliminary complete enumeration phase of the census. These two cases are commented on in the footnotes of Annex Table A6.

The above analysis shows that for the first type of combination of complete and sample enumerations, the frames used for the complete and the sample components were, in general, different while for the second and third types the same frame was used for both components of the census.

CHAPTER 5
Coverage

As indicated in the WCA 2000 Programme, an agricultural census should, in principle, cover all agricultural holdings within the entire national territory including both rural and urban areas. Overall coverage is particularly important for providing a frame for subsequent agricultural surveys. Besides, incomplete coverage leads to underestimation of census variables, and different exclusion criteria result in problems with comparability of results between countries.

In practice, however, quite often some holdings turn out not to be covered by the census. There are two types of incomplete coverage: 1) some geographic areas of the country are not covered (incomplete geographical coverage); and 2) some types of holdings are excluded (incomplete statistical, or holding type coverage). Annex Table A7 shows detailed information about both types of incomplete coverage for all reporting countries while Annex Table A8 presents the thresholds used for the census where holdings were excluded based on some minimum size criteria. Footnotes explain in detail the types of exclusion.

Table A7 shows that out of 114 reporting countries, 36 countries covered all holdings both geographically and statistically, nine countries excluded only some geographic areas, 61 countries, covering the entire territory, imposed some restrictions on holdings, while eight countries applied both geographical and holding type (statistical) restrictions. Table 5.1 below presents the distribution of these figures by regions.

Table 5.1 Number of censuses by coverage

Region	Total	Complete coverage	Incomplete coverage		
			Only Geographically	Only statistically	Both Geographically and statistically
All reporting countries and territories	114	36	9	61	8
Africa	25	14	4	5	2
America, North and Central	14	2	1	11	0
America, South	8	0	1	6	1
Asia	29	9	3	12	5
Europe	29	7	0	22	0
Oceania	9	4	0	5	0

The two subsequent sections are dedicated to the two types of coverage: geographical coverage and statistical coverage.

5.1 Restrictions on geographical coverage

As seen from Table 5.1, incomplete coverage caused by geographical restrictions is far less widespread than incomplete coverage resulting from the exclusion of some holdings: only in 17 censuses out of 114 some part of the country has been excluded from the census coverage. It should be noted that only the cases where holdings were excluded solely based on their location were counted as cases with geographical restriction. For example, South Africa excluded small-scale market gardens in peri-urban areas, but this case was counted not as geographical restriction but as an exclusion based on holding type because holdings were excluded not only for being in peri-urban areas but also for their type - small-scale market gardens. The same is true for Nicaragua which excluded kitchen gardens in urban areas. In Georgia, due to restricted financial resources, enumeration was not conducted in five large cities but this does not mean that these cities were geographically excluded because the holdings belonging to holders who lived in those cities, were spread over all the country.

The decision to exclude some parts of the country was mostly based on two reasons:

1. Agricultural production in the excluded areas was deemed to be unimportant, at least relative to resources needed for data collection there. Mostly these were urban and peri-urban areas, but often deserts and other scarcely populated or remote areas also fell into this category.
2. It was impossible to collect data on the excluded territory because it was not controlled by the government, because of adverse natural events, or for security and other reasons.

Table 5.2 describes the situation in this respect. It shows that in Europe and Oceania no geographic areas were excluded during the agricultural census, while Asia was the region with the largest number of cases of geographical exclusion. This was mostly caused by the fact that in a number of countries, part of the country's territory was occupied or not in the control of the government (Azerbaijan, Cyprus and Georgia) or data collection was not secure in some areas (Afghanistan and Myanmar). In addition to exclusion for security reasons, Myanmar, India (in some states), Saudi Arabia and Sri Lanka decided to exclude urban areas from the coverage of their agricultural censuses. India also faced some problems during the livestock data collection in some districts.

Table 5.2 Censuses excluding some geographical areas by reason of exclusion

Region	Total	Agriculture deemed unimportant	Data collection impossible	Both
TOTAL	17	8	5	4
Africa	6	4	0	2
America, North and Central	1	0	1	0
America, South	2	2	0	0
Asia	8	2	4	2
Europe	0	0	0	0
Oceania	0	0	0	0

In Africa all countries reporting geographic exclusions purposefully excluded some areas from their census coverage. Botswana, Mozambique, Senegal and Tanzania excluded urban areas. Namibia restricted its census to the northern part of the country where most of the agriculture was concentrated, while in Ethiopia pastoralist areas of Afar and Somalia Regional States were excluded from the census coverage. In addition, Mozambique had to exclude some parts of the country because of adverse natural events, while Senegal excluded some territories for security reasons.

In some censuses geographical coverage differed across census items. In the extreme northern and southern areas of Chile, data collection was carried out practically independently a year before the main census, and the main census questionnaire contained some items that were not collected in these regions. These items included information on: holder's household members and their labour input to the holding, irrigation systems, vineyards, forest plantations and age of agricultural machinery. The United States of America went even further with the introduction of 12 regional versions of the Non-sample Form and 13 regional versions of the Sample Form. These two countries, however, were not classified as countries with geographic exclusions.

The Programme noted that great care must be taken in excluding any areas in a census of agriculture. In many cases the decision was taken on the basis of the argument that the enumeration of such areas required resources which were incommensurable with their contribution to agriculture. However, the size of the contribution itself may not have been negligible, leading to serious weaknesses in the census coverage. From this point of view the example of Mozambique should be noted, where the excluded cities of Beira, Maputo, Matola and Nampula were investigated separately.

5.2 Restrictions on holdings' coverage

Over 60 percent (69 out of 114) of the countries and territories covered in this publication excluded some types of holdings during their census in the WCA 2000 round. Of them 57, a large majority, did this by establishing minimum size criteria for holdings to be included in the census. Annex Tables A7 and A8 provide detailed information about the ways used by countries for the exclusion of holdings, including precise descriptions of the minimum size criteria used. Table 5.3 summarizes this information. It shows the types of criteria used for exclusion of holdings. Utilized agricultural area, total area, area under specific crops etc., in combination with livestock numbers were the most common variables used for establishing minimum size thresholds, especially in Europe.

Table 5.3 Censuses excluding some holdings by type of exclusion

Region	Total	Holdings excluded according to							Criteria of other type
		Minimum size criteria with minimum size based on							
		Some types of area	Livestock numbers	Some types of area & livestock numbers	Value of production / sale	Value of production / sale & some types of area	Area, number of trees & livestock	Other threshold	
TOTAL	69	2	2	28	5	4	4	12	12
Africa	7	0	0	3	0			1	3
America, North and Central	11	0	0	2	2	0	4	1	2
America, South	7	1	0	2	0	0	0	3	1
Asia	17	1	2	8	0	1	0	1	4
Europe	22	0	0	12	0	3	0	5	2
Oceania	5	0	0	1	3	0	0	1	0

The column 'Other threshold' refers to the types of threshold not falling within the previous six categories. It includes: Egypt in Africa, where in addition to having a certain amount of land or a certain number of livestock, owning a piece of agricultural machinery also qualified a holding for inclusion; Trinidad and Tobago in North and Central America, where a holding had to produce 'mainly for sale' to be included in the census; Sri Lanka in Asia, where a holding was required to have a minimum area of land or produce mainly for sale; Cook Islands in Oceania, where the minimum size related to the area of garden crops or number of coconut or other tree crops was introduced. In South America into 'Other threshold' are classified Argentina (minimum size of total area or production mainly for sale), Ecuador (minimum size of agricultural land or at least one product for sale) and Venezuela (minimum size not specified). In Europe, Denmark and Finland based the exclusion threshold on agricultural area or Standard Gross Margin, Netherlands on Standard Gross Margin only, while France elaborated a complex set of minimum size criteria involving area, number of livestock and agricultural output. Estonia applied different thresholds for different types of enumeration units covered by the census: (a) agricultural holdings were defined as production units with at least 1 ha of agricultural or forest land or at least 0.3 ha of fish ponds, or with production mainly for sale; (b) agricultural households were defined as small producers mainly for own consumption with less than 1 ha of agricultural or forest land but with some minimum area of kitchen garden or minimum number of trees or livestock.

The last column of Table 5.3 covers the cases where the exclusion criteria were not based on a minimum size. In Africa, Senegal covered only holdings engaged in the rainfall (winter) agriculture, South Africa excluded small-scale market gardens in peri-urban areas, while Uganda did not include the private large-scale and institutional farm sector into the census. In North and Central America, Canada covered only the holdings producing some agricultural products for sale, while Nicaragua excluded kitchen gardens in urban areas. In South America, Brazil excluded holdings consisting of only family gardens and forest holdings. In Asia, Afghanistan and Mongolia covered only livestock holdings (in fact these were livestock censuses with no restriction on holdings), Georgia excluded holdings with holders residing in large cities, while Malaysia covered only crop holdings. In Europe, Belgium and Luxemburg used complex coverage criteria based partly on a minimum size of area and partly on other requirements like production for sale (Belgium) and being engaged in particular types of agricultural activities (Luxembourg).

There were also other cases of exclusions which did not enter into Table 5.3 Cook Islands excluded all holdings operated by institutions, communities and government, Iran did not cover the modern poultry farms, Nepal did not cover holdings operated by government organizations, corporations and other juridical persons, while Serbia excluded holdings operated by agricultural enterprises and cooperatives (legal units). All these countries introduced some thresholds as well, so they are classified in the corresponding columns of Table 5.3. The main reason for the exclusion in these cases, like in the case of Uganda mentioned above, could be the fact that information on those units was regularly available from other sources (e.g. administrative records). Another reason for the exclusions in Nepal, Serbia and Uganda could be that they derived their frame of agricultural holdings from a population census conducted prior to (in the case of Nepal), or simultaneously with (in the cases of Serbia and Uganda), the agricultural census.

The latter examples represent the cases where nothing is lost by the exclusion of some holdings because information is available from other sources. However, this may not always be so in other cases. The WCA 2000 Programme notes: 'In many countries, a minimum size limit is adopted for holdings included in the census. The rationale for this minimum size limit is that generally there are a large number of small holdings which make a very small contribution to total agricultural production but whose inclusion in the census greatly increases the workload. Although this argument is acceptable for some countries, it cannot be defended in most countries where very small holdings may contribute substantially to total agricultural production. Small holdings are often a significant part of the agricultural structure; without information on such holdings a complete picture cannot be provided. Countries that exclude small holdings from their agricultural censuses are strongly urged to set the minimum size limit as low as possible and to take steps to collect data though sample surveys for excluded holdings.' (FAO, 1995. Paragraph 4.14)

If the sample survey mentioned in the previous paragraph is a part of the enumeration process, then the census in fact is conducted with no threshold at all, with complete statistical coverage, by a combination of complete and sample enumeration methods. Box 5.1 below depicts one such example of good practice in census taking.

Box 5.1 An example of sample enumeration of holdings below the threshold

Slovakia 2001

Agricultural holdings were categorized as **above** or **below** the following threshold:

Area of utilized agricultural land of 0.5 ha; or
Area under intensive crops (orchards, vegetables and flowers) of 0.5 ha; or
Area under vineyards of 0.05 ha; or
Area under root/tuber crops of 0.03 ha; or
1 head of cattle; or
2 heads of pigs; or
4 heads of sheep or goats; or
50 heads of poultry; or
100 heads of fur-bearing animals; or
100 hares; or
5 beehives.

Short and long questionnaires were used, depending upon the size of the holding. The census was carried out as a complete enumeration of the holdings above the threshold with the long questionnaire and a sample enumeration of the holdings below the threshold with the short questionnaire.

CHAPTER 6
Scope

The scope of the agricultural census refers to the data to be collected. Consistent with the basic objective to collect structural data that change only slowly over time, the WCA 2000 Programme broadly defined the recommended scope of the agricultural census as follows:

a. Holding location.
b. Legal status of the holder.
c. Purpose of production.
d. Integration of the holding with enterprises engaged in other economic activity(ies).
e. Basic demographic characteristics of the holder and the household.
f. Inventory of production factors:
    (i) source of manpower used on the holding (family workers, hired   agricultural workers);
    (ii) number and area of land parcels;
    (iii) area by land use;
    (iv) area harvested, by crops;
    (v) number of trees by crops;
    (vi) number of livestock by type;
    (vii) type of machinery and equipment used;
    (viii) number of forest trees on the holding;
    (ix) agricultural buildings.
g. Tenure arrangements on production factors:
    (i) land tenure;
    (ii) source of machinery and equipment used.
h. Other features:
    (i) shifting cultivation;
    (ii) use of :
       - irrigation;
       - drainage;
       - fertilizers;
       - pesticides;
       - high yielding seed varieties.
    (iii) fishery or forestry activities if carried out on the holding;
    (iv) livestock system.

Based on the proposed scope, the Programme recommended a set of data items to be collected which were organized in the following ten categories:

01: Identification
02: General characteristics
03: Demographic characteristics
04: Employment
05: Land and water
06: Crops
07: Livestock
08: Machinery and equipment
09: Buildings and other structures
10: Other activities

Table 6.1 Number of censuses by coverage of census item categories

Region	Total Number of reported Censuses	Census Item Category									
		01 Identification	02 General characteristics	03 Demographic characteristics	04 Employment	05 Land and Water	06 Crops	07 Livestock	08 Machinery and equipment	09 Buildings and other structures	10 Other activities
All reporting countries	114	114	91	99	94	106	107	110	76	74	50
Africa	25	25	15	24	23	21	24	25	17	10	7
America, North and Central	14	14	12	13	11	14	13	13	9	11	8
America, South	8	8	8	6	7	8	8	8	5	6	6
Asia	29	29	21	22	19	25	25	26	16	13	12
Europe	29	29	29	27	27	29	28	29	23	28	10
Oceania	9	9	6	7	7	9	9	9	6	6	7

The data items are shown in detail in Annex Table A1, while Annex Table A9 shows country practices in covering the recommended items in national censuses. To each category of data items there corresponds a column in Annex Table A9. If a country has included in its census Programme at least one item of a category, '√' is inserted in the corresponding column. Table 6.1 summarizes by regions the information contained in Annex Table A9.

The current chapter comprises ten sections according to census item categories. Each section briefly refers to the corresponding data items and concepts of the WCA 2000 Programme and summarizes general country practices of collecting information about those items. Several examples of country practices are offered in the associated boxes.

6.1 Category 01: Identification

Under this category, the WCA 2000 Programme recommended 'holding address' and 'holder's name and address' (if different from the holding address) as essential items, while 'respondent's name' (if it was not the holder) was proposed as a non-essential item. It is important to make the distinction between the holding and the holder addresses, especially for tabulations at the lowest administrative levels.

All reported censuses collected data on at least one item from this category allowing tabulation by geographic or administrative areas. All censuses collected data about the holder's name and address. Most censuses also collected data about the holding location either directly, by asking the corresponding question, or indirectly, by collecting data on the location of parcels comprising the holding. In the latter case, two approaches were used: either the location of each parcel was recorded or information on the localities where the holding land was situated was collected. In the remaining cases, including those where the census was conducted as a module of a population census, the holding address was implicitly assumed to be the same as the holder's address. Box 6.1 presents some country examples of questions used to collect data on the identification of the holding address.

Information about the respondent was collected in less than half of all cases and even then, the name of the respondent was often not asked, information was only collected about the relation of the respondent to the holding or the holder. Knowing the respondent's name is important for controlling enumerator performance, while information on the relation of the respondent to the holder or the holding gives an indication of the reliability of the reported data.

6.2 Category 02: General characteristics

Under this category the WCA 2000 Programme identified the 'legal status of the holder' and the 'main purpose of production' (whether the holding was producing mainly for sale or mainly for own consumption) as essential items. It also recommended to collect information on 'whether the holding was part of an enterprise engaged in other economic activities' including specifying the activity if this was the case. Also, on the 'presence of a hired manager' including specifying his/her name, address and economic status (wage, salary or shares in financial returns from the holding). For legal status the Programme distinguished between two main types: private and government with further disaggregation of the former into an individual, a household, two or more individuals of different households or two or more households, corporation, cooperative, other. Hired manager referred to a person (either civil or juridical) who took technical and administrative responsibility to manage a holding on a holder's behalf, the responsibilities being limited to making day-to-day decisions to operate the holding.

Box 6.1 Some questions used to identify holding and holder addresses

---

Saint Lucia 1996
Identification of the holding
Name of holding (if any):
Location of holding:

district	locality	settlement.

Identification of the holder (in case he/she is an individual)
Name of holder ———————————————————————————
Address of Holder ———————————————————————————

Myanmar 2003

A-1 Identification of the holding:	**Geographical Location**
1. Name of Holder	1. State/Division
2. Name of Household head	2. District
3. Name of Respondent	3. Township
4. Relationship to Holder	4. Village tract/Ward
5. Name of Manager (For Juridical Holder) Address	5. Village/Block

A-4. Land Utilization and Location of Parcel

Parcel No.	Indicate Kwin Number	Enter Geographic Code of Location of Parcel		
		Towns hip	Village Tract	Village Code
(1)	(2)	(3)	(4)	(5)

Hungary 2000
1. Identification of the Holding
11. Holder's Name
12. Manager of the holding: holder (1), holder's spouse (2), other family member (3) non family member (4)
13. Address: (zip-code, town, village, district of capital)
14. Part of the settlement
15. Street, road, square, etc.

32. Area out of the settlement

No.	Settlement		Area	
	Name	Code	Hectare	m^2
a	b	c	d	

Box 6.2 Legal status data collected in selected censuses

South Africa 2000   Please indicate the ownership of the farm. Is the farm owned by:

- Individual	☐	- Private company	☐	- Government enterprise	☐
- Family	☐	- Public corporation.	☐	- Trust	☐
- Partnership	☐	- Close corporation.	☐	- Other (specify).	☐
- Public company.	☐	- Co-operative society	☐		

Guatemala 2003   Is the holder
An individual person? 1 ☐   A juridical person?   2 ☐

Chile 1997   Legal status of the holder (insert the code)   ☐
Natural person: *Juridical person:*
1 Individual producer;   3 Fiscal or municipal institution;
2 De facto society without legal contract.   4 Anonymous society or society with limited responsibility;
                                              5 Other society with legal contract (religious order,
                                                 technical school, private university etc.)
                                              6 Indigenous or agricultural (historical) community

Georgia 2004-05   Who is the holder?

Household	1 ☐	Joint Responsibility Society	4 ☐	Joint Stock Company	7 ☐
Several households	2 ☐	Comandite Society	5 ☐	Cooperative	8 ☐
Individual entrepreneur	3 ☐	Limited responsibility Society	6 ☐	Other (specify)	9 ☐

Which is the ownership form of the holding?   Public 1 ☐   Private 2 ☐

Italy 2000   2. Legal form:

2.1. Individual holding	1 ☐	*2.4. Cooperative*	10 ☐
2.2. Common land	2 ☐	*2.5. Producers Association*	11 ☐
2.3. Capital company or Partnership		*2.6. Public body*	
a) Simple	3 ☐	a) State	12 ☐
b) Partnership	4 ☐	b) Region	13 ☐
c) Limited Company	5 ☐	c) Province	14 ☐
d) Limited partnership	6 ☐	d) Municipality	15 ☐
e) Partnership limited by shares	7 ☐	e) Mountain Community	16 ☐
f) Joint-stock Company	8 ☐	f) Other (specify)	17 ☐
g) Other type (specify)	9 ☐	*2.7. Other legal form*	
		a) Consortium (specify)	18 ☐
		b) Other type (specify)	19 ☐

Guam 2002   What type of operating organization does this place have?

1 ☐	Individual;	3 ☐	Corporation (do not include cooperatives);
2 ☐	Partnership;	4 ☐	Other (Cooperative, estate, trust, etc.).

As seen from Table 6.1, up to 80 percent of the censuses collected at least one item from the 'general characteristics' category. Practically all of them included legal status of the holder. However, quite often legal status types were not in line with those recommended by the Programme. Even the wording of the corresponding question did not always contain the expression 'legal status'. Box 6.2 depicts several examples of country practices of data collection related to this item.

Less than half of the reporting countries collected information on hired manager as the person who makes day-to-day decisions on the holding. In most cases only the information about whether there was a hired manager was collected without asking his/her name or address. Information about whether there was a hired manager was obtained either as a response to a direct question or by asking questions on the structure of the paid employment on the holding.

About half the censuses collected information on other economic activities of the holding enterprise. In practically all cases the census questionnaire did not ask whether the holding was a part of an enterprise also engaged in other economic activities – information on other economic activities was obtained through questions on other gainful activities the holding was engaged in or if there was another source of income for the holding. However, in the case of holdings operated by households this was equivalent to asking whether the holding was part of an enterprise managed by the household as an institutional unit engaged in production.

Finally, in about half the cases the collected data made it possible to identify the main purpose of production, whether it was for mainly for sale or own consumption. Sometimes this was achieved by asking a direct question, sometimes by asking this question for each crop and livestock product produced by the holding, and sometimes by analyzing production and sales data of the holding (a number of censuses collected data related to production and sales even though these items were not included in the WCA 2000 Programme).

6.3 Category 03: Demographic characteristics

This category applies to the holdings belonging to the household sector (holdings operated by one individual, a household, two or more individuals of different households or two or more households), and describes the holder and holder's household members. The proposed items were: number of household members, and for each of them (including the holder) name, age, sex, marital status and education. Age and sex of the household members were recommended as essential items.

It can be seen from Table 6.1 that the vast majority of countries collected some type of demographic information. In the countries where the agricultural census was undertaken simultaneously with the population census, demographic data were collected in the population census questionnaire with even more detail than recommended by the WCA 2000 Programme. These cases were also included in Tables A9 and 6.1 under the heading 'Demographic characteristics'. Chapter ten reviews in more detail the cases of integration of agricultural and population censuses.

Country practices of collecting demographic data vary considerably. Most of countries collected information about the size of holder's household. However, not all of them collected individual data for each household member. Some countries, like Azerbaijan, Chile and Georgia, collected only the number of household members in certain age and sex groupings. Some countries collected individual demographic information only for the holder, while some countries collected individual demographic information only for household members above a certain age threshold. Even when individual demographic information was collected, not all recommended items were covered. This is especially true for the item 'marital status' which was included in only a few census questionnaires. Box 6.3 provides examples of all types. Because individual demographic data are often collected together with employment data, some examples contain employment data as well.

Box 6.3 Some examples of collecting demographic and employment information on household members

Tanzania 2004 Give details of personal ***particulars*** of all household members beginning with the head of the household

Names of household members	Relationship to head	Sex M=1 F=2	Age	Survival of parents		Read & Write	Education Status	Edication level reached	Involvement in farming	Main activity	Off-farm income Yes=1 No=2
				Mother	Father						
(1)	(2)	(3)	(4)	(5)	(6)	(7)	(8)	(9)	(10)	(11)	(12)

Not applicable for children under 5 years of age (applies to columns 7–12)

Codes. (Col.2): Head of household=1; Spouse=2; Son/Daughter=3; Father/Mother=4; Grandson/Granddaughter=5; Other relative=6; Others=8. **(Col 5,6):** Yes=1; No=2; Don't know=3. **(Col 7):** Swahili=1; English=2; Swahili & English=3; Any other language=4; Don't read/write=5. **(Col 8):** Attending school=1; Completed=2; Never attended school=3. **(Col 9):** Not of school age=NA; Under standard One=00, Standard One=01; ... Standard Eight=08; Training after Primary education=09; PreForm One=10; Form One=11; ... Form Six=16; Training after Secondary Education=17; University and other tertiary education=18; Adult Education=19; Not applicable=99. **(Col.10):** Works full time on farm=1; Works part-time on farm=2; Rarely works on farm=3; Never works on farm=4. **(Col.11):** Crop farming=01; Livestock keeping/Herding=02; Livestock pastoralism=3; Fishing=4; Paid employment - Government parastatal=05; Private/NGO/Mission/etc=06; Self employed (non farming) – with employees=07; without employees=08; Unpaid family helper (non agriculture)=09; Not working & available=10; Not working & unavailable=11; Housemaker/housewife=12; Student=13; Unable to work/too old/retired/sick/disabled=14; Other=98.

Azerbaijan 2005 Labour force in agriculture (part of the table)

Indicators	Row #	Total	Including	
			Male	Female
Total number of family members in peasant (farmer) farms				
Including children under 15				
Persons at able-bodied age: men 15-62, women 15-57 years old				
Of which youths aged 16-35				
Persons over able-bodied age: men over 62, women over 57 years				

Question 7. Sex of enterprise head (indicate by X the corresponding answer)
7.1. Male ❑   7.2. Female ❑   7.3 Age ❑ ❑

Question 8. Professional education of enterprise head (indicate by X the corresponding answer)
8.1.Agricultural education (higher, secondary) ❑   8.2.Incomplete agricultural education (higher, secondary) ❑
8.3.Non-agricultural education ❑   8.4. No professional education ❑

Northern Mariana Islands 2002 The following questions are about the CHARACTERISTICS of the FARM OPERATOR.
1. RESIDENCE – Does the operator live on this place?   1 ❑ Yes   2 ❑ No
2. In what YEAR did the operator begin to operate any part of this place? _____Year
3. AGE of the operator on last birthday _____Age
4. PRINCIPAL OCCUPATION – At which occupation did the operator spend the majority (50 percent or more) of his/her worktime in 2002?   1 ❑ Farming or ranching   2 ❑ Other
5. OFF-FARM WORK – How many days did the operator work at least 4 hours per day off this place in 2002? Include work at a nonfarm job, business, or on someone else's farm (exclude exchange farm work)
    1 ❑ None   2 ❑ 1-49 days   3 ❑ 50-99 days   4 ❑ 100-149 days   5 ❑ 150-199 days   6 ❑ 200 days and more
6. RACE/NATIONAL ORIGIN of the operator
    1 ❑ Chamorro 2 ❑ Chinese 3 ❑ Filipino   4 ❑ Japanese 5 ❑ White   6 ❑ Other-Specify
7. SEX of the operator   1 ❑ Male   2 ❑ Female

6.4 Category 04: Employment

This category was designed to provide a minimum data set on employment on the holding. For each household member associated with a holding, the items 'whether economically active', 'whether has more than one occupation' and 'type of main occupation' were proposed as essential items while 'whether any work done on holding during the year' and 'whether permanent or occasional worker on holding' were recommended items. The Programme also recommended collecting some information about agricultural workers other than members of the holder's household, specifically: whether permanent/occasional agricultural workers employed during the year and number of permanent agricultural workers – male/female.

Most of the reported censuses (94 out of 114) collected data for at least one item from the employment category (see Table 6.1). Practically all of the countries collected employment data on holder's household members. These were generally collected along with information on demographic characteristics. For censuses conducted jointly with the population census, the employment data of household members were collected in the population census questionnaire. Country practices show different approaches to the collection of employment data of household members. Some countries collected information only for holders. Some countries placed emphasis on the work done by household members on the holding and were not concerned with the specific occupations of the household members. When collecting individual data about work on the holding, several practices were used: (a) asking only about whether the household member did any work on the holding, not being interested in the permanent or temporary character of the work; (b) asking directly whether the household member had worked permanently of temporarily on the holding with some criteria to distinguish between the two (the criteria varied across countries); (c) asking number of days per year or average hours per week/per day worked, making it possible to analyse the permanent or temporary nature of the work.

Data on employment of agricultural workers other than the holder's household members, were collected by a lesser number of countries than data on employment of holder's household members. They were not collected by the countries that conducted the agricultural census in conjunction with the population census. The extent of information collected varied from country to country. Most countries in the Africa region covered this topic according to the recommendations of the WCA 2000 Programme, asking whether permanent and temporary workers were employed during the year and the number of permanent workers broken down by sex. Many countries, however, especially in Europe, went beyond the Programme recommendations and asked about days or hours worked by permanent employees, in some cases collecting information even for each permanently employed worker.

Box 6.3 contains some examples of country practices of collecting household members' employment data, while Box 6.4 shows some country examples of collecting employment data of agricultural workers other than household members.

Box 6.4 Some examples of collecting employment data for non-household members

Mali 2004-05

4.13 - Utilization of the hired labour force
Do you use hired labour force on the parcel? (Yes=1, No=2) _____
If yes, what is the nature of the labour force _____
1=permanent worker; 2=temporary worker; 4=worker to a task; 8=mutual aid group;
16=other (associated)

4.14 - Job performed by the hired labour force _____
1= clearing   2=soil preparation;   4=seedlings/transplantation;   8=weeding;
16=spreading and treatment;   32=harvesting;   64=shuffling/winnowing;   128=transportation

Puerto Rico 2002   SECTION 25. HIRED WORKERS, AGREGADOS, AND SHARECROPPPERS
In 2002, did you hire any workers to do farmwork on 'THIS OPERATION' or was any agregado family or sharecropper living here on December 31, 2002? (Exclude employees of labour contractors who did work for you)

❑ Yes – complete this SECTION   ❑ No – go to SECTION 26   None   Number
1. How many of the hired workers were employed on 'this operation' for 5 months or more?   ❑   _____
2. How many of the hired workers were employed on 'this operation' for less than 5 months?   ❑   _____
3. How many agregados or sharecropper families were employed on 'this operation' on
    December 31, 2002?   ❑   _____

Estonia 2001
Regular employees (except sole holder's members of the family)

Row #	Age group	Working time in the agricultural holding (except housework), % of the full wo king time of one worker in a year)			
		0-<25	26-<50	50-<100	100
Males, number					
222	15-19				
...	...	...	...	...	...
232	65-74				
233	Over 74				
Females, number					
234	15-19				
...	...	...	...	...	
244	65-74				
245	Over 74				

Temporary employees

246		Number		Hours worked

6.5 Category 05: Land and water

As stated in the WCA 2000 Programme, this category provided basic data on the holding's land and water resources. The items were divided into two classes: those to be collected for the entire holding and those to be collected for each parcel. It was recognized, however, that some countries might prefer to collect all data at the holding level.

At the holding level, the 'number of parcels' and the 'total area of the holding' were proposed as essential items, while 'whether land was rented to others', 'area of land rented to others' and 'relative area of soil degradation of the holding' were proposed as non-essential items.

All other items of this category were recommended to be collected at the parcel level. At the parcel level, essential items were: total area of the parcel, land tenure, land use, whether irrigated some time during the year, and whether shifting cultivation practised. Non-essential items referring to the parcel were: its location, whether drainage facilities were available, area irrigated, area affected by salty soil or high water table, area with irrigation potential, and items referring to soil (soil type, soil colour, soil depth, soil salinity, surface drainage, rate of percolation, soil degradation level). Wherever shifting cultivation was relevant, the Programme also recommended to collect information about the year when the parcel was cleared for cultivation.

Practically all reporting countries collected data for at least one item from Category 05 (among the countries who did not do so were Afghanistan and Mongolia which conducted only livestock censuses). All of them except Columbia collected data about the total area of the holding. Columbia, using area sampling, collected land data for plots of sample segments ('pedazo de segmento de muestreo'), not for holdings. In some cases, instead of the total area of the holding, information was collected about utilized agricultural area or cultivated area.

About two thirds of the countries collected data about land fragmentation. Most of them did it in accordance with recommendations of the WCA 2000 Programme, that is, they asked about the number of parcels comprising the holding, a parcel being understood to be any piece of land entirely surrounded by water or land not forming part of the holding. In other cases information was collected about the number of different localities where the holding land was situated.

Very few countries collected information about land rented out. This may have been due to the fact that land rented out was not part of the holding land, so this information was rather about the holder than the holding which was the enumeration unit of the agricultural census.

Land use is one of the most important items to be collected in the agricultural census. The Programme distinguished between six major land use classes into which the total land of the holding should be disaggregated:

1. Arable land
2. Land under protective cover
3. Land under permanent crops
4. Land under permanent meadows and pastures
5. Woodland or forest
6. All other land

More than half of the countries in the Africa region did not collect land use data, focusing more on crops grown on the land rather than land use categories. In all other regions the overwhelming majority of the countries included a breakdown by land use in their questionnaires. In many cases, however, the land use categories used by countries were not in complete conformity with the categories recommended by the WCA 2000 Programme. Box 6.5 gives some examples of land use classes used by various countries. The majority of the countries preferred to collect land use data for the whole holding instead of doing so for each parcel, as recommended by the WCA 2000 Programme.

Box 6.5 Some examples of land use classes used by countries

Canada 2001
45.Total area of hay and field crops
72.Total area of vegetables
73.Total area of nursery products
74.Total area of sod grown for sale
89.Total area of fruit, berries and nuts
90.Total area of Christmas trees grown for sale
91. Total of questions 45,72,73,74,89,90
92. Summerfallow
93. Tame or seeded pasture
94. Natural land for pasture
95. All other land on this operation
96. TOTAL land area (Total of questions 91 to 95)

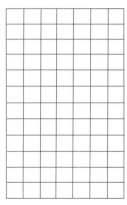

Colombia 2001 Land use categories:
Temporary crops; Permanent crops; Fallow land; Land under rest; Pastures and meadows; Weeds and stubbles; Natural forests; Planted forests; Porciculture; Aviculture; Floriculture; Pisciculture; Wasteland and rocks; Water bodies; Other uses.

Japan 2000 Land use categories:
1   Paddy fields;
2   Land under permanent crops (excluding ordinary upland fields and short time meadows);
3   Upland fields;
4   The total area of cultivated fields you are managing ((1)+(2)+(3));
5   Land used as meadows and grazing land, excluding cultivated land (Please fill in areas of land used in forests and wilderness in the past year);

Lithuania 2003 Land use categories:
100. Kitchen gardens; 101. Arable land; 102. Pastures and meadows (5 years or more);103. Rough grazing; 104. Fruit and berry plantations; 105. Nurseries; 106. Other permanent crops (oster, reeds etc.); 107. Marketable greenhouses; 108. Utilized agricultural area (sum of 100-107); 109. Unutilized agricultural land; 110. Wooded area; 111. Water; 112. Of which water bodies with fish; 113. Other land (land occupied by buildings, tracks, marshland and other non-agricultural land); 114. Total land area (sum of 108-111, 113)

Land tenure is another important census data item. Land tenure categories according to the WCA 2000 Programme were as follows:

   a.   Area owned or held in owner-like possession
   b.   Area rented from others
   c.   Area operated on a squatter basis
   d.   Area operated under tribal or traditional forms of tenure
   e.   Area operated under other forms of tenure

The overwhelming majority of countries in each region collected data about land tenure during the WCA 2000 round. However, very few of them collected the data by parcel as recommended by the Programme. In addition, land tenure classifications used by countries varied considerably, many being different from those recommended. Some countries tried to introduce their specific land tenure

Box 6.6 Some examples of land tenure classes used by countries

Senegal 1998-99 Mode of tenure or acquisition: Indicate the mode of tenure of the parcel ...... [   ]

Codes: 1=Heritage; 2=Purchase; 3=Rent or sharecropping; 4=Loan; 5=Gift; 6=Other

Canada 2001   TOTAL area of land:

15. **Owned** (Do not include land rented to others)..
16. **Leased FROM governments** (include land operated under licence, permit or lease, etc.)
17. **Rented or leased FROM others** ....
18. **Crop shared from others**..
19. **TOTAL area of land of this operation** (Total of questions from 15 to 18).....................

Cyprus 2003-04   **6.1 How much of the holding's utilized agricultural area is:**   [Decares]

a) **Owned**, i.e. belongs to the holder or to members of his family... [   ]
b) **Rented** for a specific amount (in kind or money) from other owners....................................... [   ]
c) **Jointly owned,** i.e. part of the produce is taken by the owner of the land and part by holder in accordance with an agreement. *Jointly owned agricultural land should be declared by the person who works the land, not by the owner*.. [   ]
d) **Agricultural land with different exploitation status,** i.e. coded for no rent, farmed arbitrarily, etc [   ]

Slovenia 2000   **C. LAND**

		Hectares	Ares
01	Land owned		
02	Land rented from others		
03	Land leased to others		
04	TOTAL UTILIZED AREA (01+02-03)		

classifications. Some of them restricted themselves only to owned and rented land, grouping all other types into the category 'other' or even not mentioning them at all. Box 6.6. highlights some country practices in this respect.

Most countries collected some data about irrigation. In the Africa region most of the countries collected irrigation data for each parcel, as recommended by the WCA 2000 Programme, while in other regions countries mostly preferred to collect information for the whole holding. In practically all the cases of collection of data on irrigation, information on area irrigated was collected, quite often for each crop. As to information about the irrigation potential of the holding (irrespective of whether or not it was actually irrigated during the reference period), it was included by some 25 percent of the countries that collected irrigation data.

Less than a quarter of the countries collected information about the existence of drainage facilities. Where collected, this information was in most cases for the holding as a whole.

Practically no country collected information about shifting cultivation and soil. In this respect, Lao People's Democratic Republic and Nepal should be mentioned. The former included a question about how many years ago had the parcel been cleared, while the latter collected information about the type and colour of the soil.

6.6 Category 06: Crops

This category covered the basic features of crop cultivation on the holding. Items were grouped in five main categories, namely: temporary crops, permanent crops, fertilizers, pesticides and seeds.

For temporary crops, two items were proposed, both as essential items: the name of the crop grown and the crop area harvested.

For permanent crops, the number and area under trees of productive age in compact plantations were recommended as essential items, while names of the crops grown, number of scattered trees and area under trees of non-productive age in compact plantations were recommended as non-essential items.

For fertilizers, whether inorganic fertilizers were applied was proposed as an essential item while whether organic manure or other fertilizers were applied was proposed as a non-essential item. The amount of inorganic fertilizers applied per crop was also recommended as a non-essential item.

For pesticides, whether pesticides were applied during the year was proposed as an essential item while the frequency of pesticide applications per crop was recommended as a non-essential item.

For seeds the Programme proposed whether high yield variety seeds were used during the year as an essential item while collecting information about specific crops with high yield and traditional varieties of seeds was considered as non-essential.

As Table 6.1 shows, nearly all countries collected information on at least one item from Category 06. All of them collected information on temporary crops. As to permanent crops, Namibia did not collect that information because it restricted its agricultural census to collecting data only for pearl millet ('mahangu', as it is called in Namibia), sorghum and maize. Some countries, conducting the census jointly with the population census, collected information only on the names of crops grown.

For temporary crops, nearly all countries collected information on the area for each crop. In some cases this was the harvested area as recommended by the Programme, in other cases this was the sown area, and in yet some other cases both harvested and sown areas were collected, depending on the time when the enumeration took place. Sometimes the census questionnaire asked about the area of the temporary crops present at the enumeration moment. The United Kingdom and Belgium even asked about areas intended to be sown during the current agricultural season.

Box 6.7 provides some country practices used for collecting data on temporary crops.

For permanent crops country practices were even more diverse. Above 80 percent of countries collected information on areas of crops in compact plantations. Practically all reported censuses in the Europe, North and Central America and South America regions did so, while countries in the Oceania region placed more emphasis on the number of trees than areas of compact plantations. In most cases countries did not differentiate between areas under trees of productive and non-productive age.

Box 6.7 Country practices of data collection on temporary crops

Mali 2004-05 **10-SUMMARY OF BLOCS AND PARCELS OF THE HOLDING**

Bloc #	Parcel # within the bloc	Season 1=Winter 2=Not winter	Main crop		Area in sq. m.	Ordinal # of the person responsible for the parcel	Name of the person responsible for the parcel
			Name	Code			

Saint Lucia 1996 State the following information for TEMPORARY crops present in the holding

TEMPORARY CROPS	IS THIS CROP GROWING TODAY?		WAS IT GROWN DURING 1995?		IF THE ANSWER TO ANY TWO OF THE PREVIOUS QUESTIONS WAS YES STATE THE END USE OF THE CROP		
	YES	NO	YES	NO	HOME USE	HOME USE/SALE	PRIMARILY FOR SALE
1. TANNIA	1	2	1	2	2	3	
...	...	...	...	...	...	...	...
27.RADISH	1	2	1	2	1	2	3
OTHER (Specify): CODE							
	1	2	1	2	1	2	3

Belgium 2003 **4. MAIN CROPS**

- Provide information on crops present as of 1 May
- Do not forget the crops under contract
- For uncultivated land indicate the main culture which will be sown during the current cropping season (for vegetables provide data only for the first crop which will be sown after 1 May)

4.1. Areas permanently under grasses (include old grassland renovated after 1 May 1998)

		Code	Hectares	Ares
Grassland sown before	Pasture	001	....	....
1 May 1998 mainly for	Mowing	002	....	....
	TOTAL	009	....	....

4.12. Seedlings and saplings in open air

	Code	Hectares	Ares
Seedlings for vegetables	181	....	....
Seedlings and saplings for ornamental crops	182	....	....
TOTAL	189	....	....

Box 6.8 Country practices of data collection on permanent crops

Georgia 2004-05

8. Numbers of fruit and citrus trees							9. Numbers of vines by species				
Type	#	Number of trees in compact plantations		Number of scattered trees		Vine species	Code	Number of vines in compact plantations		Number of scattered vines	
		Total	Of which of producing age	Total	Of which of producing age			Total	Of which of producing age	Total	Of which of producing age
Apple	1					Rkatsiteli	1				
...	.	.	...	.	...	...	.	.	...	.	...
Apricot	6					Izabela	6				
...	.	.	...	.	...						
Grapefruit	21										

Chile 1997

Section III: Vineyards					Section IV: Orchards			
Groups of Varieties	Area (hectares)				Fruit varieties	Area of compact plantations (Hectares)		
	Not irrigated	Irrigated						
	Vineyards (viñas)	Vineyards (viñas)	Vineyards (parronales)				Growing	Producing
Domestic 112					Almond 117			
Red 113					Cranberry 118			
White 114	...	...			...			
Pisqueras 115					Fruit nursery 142			
TOTAL 116					TOTAL 143			

Northern Mariana Islands 2003   SECTION 4

Were any FRUITS, NUTS, or NURSERY CROPS grown or harvested FOR SALE in 2002?
1 ☐   YES –Complete this section   2 ☐   NO – Go to section 5

Whole acres	Tenths
	/10

	None	How many trees or plants are not of bearing age?	How many trees or plants are of bearing age?	How many pounds were harvested in the last 12 months?
1.Avocados	☐			
...	...	...	...	...
18.Nursery crops	☐			

Data about the number of trees were collected by less than two thirds of the countries. Some countries collected data only on the number of scattered trees, some only on the number of trees in compact plantations and some collected both types of information, while other countries collected data about the total number of trees by crop, without distinguishing between scattered trees and trees in compact plantations.

Box 6.8 provides some country practices used for collecting data on permanent crops

Quite a number of countries (Albania, American Samoa, Australia, Botswana, Chile, Columbia, Comoros, Guam, Guatemala, Madagascar, Malta, Mozambique, Namibia, New Zealand, Northern Mariana Islands, Panama, Puerto Rica, Slovakia, South Africa, Tanzania, United States of America and US Virgin Islands) collected information on crop production. Production (or yield) was not considered as an item in the WCA 2000 Programme because it fell outside the basic census criteria of being a structural item only changing slowly over time. Its collection by countries may, however, be attributed to various reasons. In some cases the system of current agricultural surveys was not well established and the agricultural census was seen as the only opportunity to obtain production and yield data. In other cases countries wished to have a benchmark to compare with the results of the current agricultural surveys.

Less than 75 percent of the censuses collected information on the use of fertilizers. Very few of them asked about the amount used per crop, as recommended by the Programme. Instead, a number of countries collected information about areas fertilized. Here country practices varied significantly. Some countries asked about areas fertilized per crop, some countries asked only about crops that were treated by fertilizers without inquiring about areas, while some countries asked only about the total area fertilized. Some countries, including American Samoa, Guam, Northern Mariana Islands, South Africa and the US Virgin Islands, obtained information on fertilizer use indirectly, by asking, within the general context of farming expenditures, about the value of fertilizers and manure purchased.

Less than two thirds of the censuses collected information on use of pesticides. Practically none of them asked about the frequency of application per crop, as recommended by the WCA 2000 Programme. Some countries asked about the crops treated with pesticides whilst others asked about the total area treated and others about the area treated per crop.

Only about 20 percent of countries collected data about the use of high yield varieties of seeds. Practically all of these countries did this per crop, in accordance with the WCA 2000 Programme recommendations. Africa and Asia were the most active regions in collecting this type of information with some two thirds of African countries and half the Asian countries collecting such data.

6.7 Category 07: Livestock

This category referred to all animals forming part of the holding, kept or reared mainly for agriculture purposes. Essential items were: type of livestock production system (nomadic or totally pastoral, semi-nomadic or semi-pastoral, sedentary pastoral, ranching) and animal numbers by type, age, sex and purpose for each relevant kind of livestock: cattle, buffaloes, sheep, goats, pigs, horses, camels, mules and hinnies, asses and chickens.

Non-essential items were those referring to 'Poultry other than chickens' (ducks, geese, turkeys, guinea fowls, pigeons and other poultry) as well other domesticated animals (beehives and bee colonies, rabbits and hares, lamas and alpacas and fur-bearing animals).

Practically all countries collected information on their livestock population. They adapted the types of livestock on which to collect data to their specific conditions. The characteristics of livestock that were covered by the censuses also varied depending on the livestock type and the country.

Box 6.9 shows some country practices of livestock data collection.

Box 6.9 Country practices of livestock data collection

Comoros 2004

Livestock type	Code	Number	Possession status	Destination
Bovines				
Sheep				
Goats				
Asses				
Chicken				
Ducks				
Rabbits				
Beehives				
Other				

Possession status codes:1=owned; 2=sharing/guardian; 3=other
Destination codes: 1=drought animal; 2=for meet; 3=for milk; 4=other

United States of America 2002   SECTION 9 (HOGS AND PIGS)

Did YOU or ANYONE ELSE have any hogs or pigs on **'THIS OPERATION'** during 2002?

1 ❏   YES –Complete this section   2 ❏   NO – Go to section 10

31 DECEMBER 2002 INVENTORY

1.   Of the total number of hogs and pigs on hand, how many were -

	None	NUMBER ON THIS OPERATION 31 DECEMBER 2002
a. HOGS and PIGS used or to be used for BREEDING?	❏	....
b. Other hog and pigs, including market hogs?	❏	....

2.   TOTAL HOGS AND PIGS on hand 31 December 2002 –
ADD items 1a and 1b   ....

HOGS AND PIGS SOLD FROM 'THIS OPERATION' IN 2002   NUMBER

3.   Number of HOGS and PIGS sold or moved from 'THIS OPERATION'
in 2002, including *feeder* pigs.   ❏   ....

China 1997   V. Livestock and Poultry (heads)

5.1 Livestock	Code	Number on hand on 31 Dec. 1996	Of which: Drought animals
Cattle	501		
Buffalos	502		
Milk cows	503		
Yaks	504		
Horses	505		
Donkeys	506		
Mules	507		
Camels	508		

5.2 Slaughtered and sold beef cattle during 1996 (509) _____ head(s)

5.3 Goats, sheep and hogs	Code	Number on hand on 31 Dec. 1996	Slaughtered and sold
Goats	510		
Sheep	511		
Hogs	512		
Of which: reproductive sows	513		
5.4 Poultry			
Chickens	514		
Meat chickens	515		
Ducks	516		
Meat ducks	517		
Geese	518		
Meat geese	519		

6.8 Category 08: Machinery and equipment

Category 08 identifies machinery and equipment used by the holding, wholly or partly, for agricultural production. There were no essential items proposed in this category. Non-essential items were: number of stationary power producing machinery used on the holding by type, and whether other machinery and equipment were used on the holding by main source and type. Proposed source classes were: owned solely by holder, owned jointly by holder and others, provided by landlord, provided by other private holdings excluding cooperatives, provided by cooperatives, provided by a private agricultural service establishment, provided by a government agency.

Box 6.10 Country practices of data collection on Machinery and equipment

Botswana 2004

How many of the following do you own and are either in good working condition or temporarily out of order?

	Working condition				Working condition		
	Good	Temporarily out of order	Total		Good	Temporarily out of order	Total
1.Harrow				14.Traller			
2. Single plough				15.Cart			
3.Double plough				16.Shovel			
4.Single planter				17.Spray race			
5.Double planter				18.Spray pump			
6.Cultivator				19.Dpi-tank			
7.Mpwer				20.Maize grinder			
8.Hoe				21.Maize sheller			
9.Rake				22.Bailer			
10.Wheel barrow				23.Forage Harvester			
11. Thresher				24.Double harvester			
12.Tractor				25. ---------------------			
13.Truck				26. ---------------------			

Albania 1998

23. Mechanical equipments used in agricultural year 1997-1998	Code	Owned only by holder			In joint ownership		Provided by the thirds
		Total Number	Used in		Own holding	Out of ownership	thirds
			Own holding	Other holding			
	1	2	3	4	5	6	7
			1=yes; 2=no				
23.1 TRACTORS							
a)Up to 15 CV	01		—	—	—	—	
b)Over 15CV	02		—	—	—	—	
Total (01+02)	03						
23.2 Auticombine	04		—	—	—	—	
23.3 Threshing machine	05		—	—	—	—	
23.4 Vehicles for commodities							
-up to 1.5 tons	06		—	—	—	—	
-over 1.5 tons	07		—	—	—	—	
Total (06+07)	08		—	—	—	—	
Other mechanical equipment; 1=yes 2=no	09	—					

As shown by Table 6.1, some 76 countries out of 114 reported having collected machinery and equipment data. However, country practices varied considerably with respect to the type of information collected. About half of these countries collected information in accordance with the Programme recommendations, that is, asking about the use of machinery with an indication of the source. A number of countries also collected information about the number of items of machinery and equipment owned. The classes used for the source of machinery and equipment used varied across the countries and did not always coincide with those recommended by the Programme. Other countries collected information only about the machinery owned or present on the holding. Box 6.10 presents some examples of country practices in this respect.

6.9 Category 09: Buildings and other structures

This category was concerned with information on the use of non-residential buildings on the holding, such as warehouses, stores, stables, barns, office buildings etc. There were no essential items in this category. The proposed non-essential items were: whether any non-residential building used, tenure for each non-residential building used, area or volume depending on purpose of use.

About two thirds of the reporting countries collected at least one item from this category. In most of these cases only the 'use' or 'presence' of agricultural buildings was collected. Quite a few censuses, with the notable exception of the Europe region, collected data on the capacity of buildings. Even fewer censuses inquired about the tenure of the buildings. Country practices concerning types of buildings about which the information was collected varied considerably, reflecting the various methods of land cultivation and livestock keeping.

Box 6.11 presents some examples of country practices in this respect.

6.10 Category 10: Other activities

This category identified holdings carrying out forestry, fisheries and other activities simultaneously with agricultural activities. Items in this category were grouped into three groups: forestry, fisheries and other activities.

For forestry, all proposed items were non-essential, specifically: whether forest trees existed on the holding, the total area of forest trees, the age of forest trees, area reforested in the last five years, whether forest products were harvested or not, value of sales of forestry products.

For fisheries, the item 'whether fish or other aquatic animals and plants were taken from the water within the holding' was proposed as a non-essential item. However, the presence of an aquaculture installation on the holding with an indication of its type, kind of product and value of sales were considered essential items by the Programme.

The Programme also recommended to collect information about other activities carried out on the holding.

As seen from Table 6.1, only 50 countries out of the 114 reporting collected data from this category. Practically all countries which included some items about forestry and/or fisheries went beyond the recommendations of the WCA 2000 Programme and incorporated elements of forestry and aquaculture/fishing censuses, respectively. Chapter ten discusses these issues in more detail.

Box 6.11 Country practices of data collection on buildings and structures

Puerto Rico 2002

How many of the following were on 'THIS OPERATION' on 31 December 2002 (include all machinery, equipment, and facilities on the operation, regardless of ownership, provided it was in working order)

b. BUILDINGS

	None	NUMBER
(1) Buildings used to house livestock..	❏	....
(2) Storage buildings for crops...	❏	....
(3) Buildings for machinery...	❏	....
(4) Greenhouses...	❏	....
(5) Houses for agregados and other workers................................	❏	....

Hungary 2000 **BUILDINGS AND STRUCTURES, 2000**

No.	Denomination (unit of measurement of capacity)			Buildings used		Buildings owned	
				By the holding			
				Number	Total capacity	Number	Total capacity
				During the last year		On 31 March, 2000	
a	b			c	d	e	f
01	Cowshed, heads						
...	...			...	...	...	...
06	Solid dung	Storage	M²				
07	Liquid	Storage	M³				
08	Slurry	Storage	M³				
...	...			...	...	...	...
25	Well (of any kind)						
...	...			...	...	...	...
29	Storage for sewage, M 3						
...	...			...	...	...	...
30	Technical Total (from 01 to 29)						

CHAPTER 7
Data processing and tabulation

The data collected in an agricultural census must be processed, that is, checked, edited, coded, captured, verified and tabulated. Most data processing operations involve both manual and electronic data processing with the manual processing being limited, in most cases, to the checking, editing and coding the forms prior to data entry or data capture. This manual processing can also be used to generate some preliminary census results.

The present chapter deals with data entry/capture, machine checking, editing and verification, and tabulation. It consists of two sections. In the first section practices and computer Programmes used by countries for data entry and for processing and tabulation are briefly reviewed. The second section deals with country practices in the use of size classes for key classification variables compared with the recommendations of the WCA 2000 Programme.

7.1 Data entry and processing

Data entry refers to the procedure used to capture information obtained during the census in a format that can be interpreted by a computer. Although it represents only a small part of the agricultural census activities, it is deemed to be one of the most critical aspects in terms of cost and time. Rapid advances in data entry technology have greatly increased the speed of database creation and database accuracy and reliability.

There are a variety of methods of data entry in census projects. They include:

- Manual data entry (key punching);
- Scanning (including laser copy process, optical character recognition (OCR) systems, intelligent character recognition (ICR) systems and optical mark recognition (OMR) systems);
- Data collection and entry through computer-assisted telephone interviews (CATI);
- Data collection through the internet.

In the WCA 2000 round, all reporting countries used computer facilities for data entry, consistency checking and the tabulation of their census data. Nevertheless, manual processing was still common at some stages of the treatment of the census data, for example the checking, editing and coding of the census forms, prior to data entry, and for producing preliminary results.

As pointed out in Section 3.3, most data were collected through hard copy questionnaires either via face-to-face or telephone interviews, or by postal enquiries. They were then entered into the computer either by scanning or using a manual data entry Programme. Only a few countries reported on the computer Programmes used for entering data and for the processing of census tables.

Most countries developed their own specific software for data entry and/or data processing, designed according to the questionnaire and tabulation plan of the agricultural census.

Regarding the use of standard software packages, the most commonly used one for data entry was IMPS (Integrated Microcomputer Processing System) and its successor CsPro (Census and Survey Processing System), while for data processing and tabulation, SPSS (Statistical Package for Social Sciences) was the most commonly used software.

Tables 7.1 and 7.2 below show the countries reporting on their computer software used for data entry and data processing and tabulation.

Table 7.1 Reporting countries by main computer software used for data entry

Region	Software used for manual data entry					Scanning**
	IMPS/CsPro		SPSS	Other*		
Africa	Cape Verde Ethiopia Madagascar Mozambique Senegal	Côte d'Ivoire Guinea Mali Namibia	Gambia	Botswana Morocco Togo	Lesotho Seychelles	Tanzania
America, North and Central				Guatemala Puerto Rico Saint Lucia St. Vincent and the Grenadines		Canada United States of America
America, South				Chile Colombia Ecuador		Argentina Uruguay
Asia	Myanmar			Afghanistan Georgia Saudi Arabia		China
Europe			Latvia	Albania Denmark Lithuania Slovenia	Belgium Italy Romania	Norway
Oceania	Samoa			American Samoa Australia Guam		

* Most of these Programmes were specially designed for the census purpose
** In this category are included: laser copy process, OCR, ICR and OMR systems.

Table 7.2 Reporting countries by main computer software used for data processing and tabulation

Region	Software used for data processing and tabulation				
	IMPS/CsPro	SPSS	SAS		Other
Africa	Cape Verde Ethiopia Namibia	Côte d'Ivoire Guinea Mali Senegal Tanzania	Gambia Lesotho Madagascar Seychelles	Botswana Togo	Mozambique
America, North and Central				Canada United States of America Saint Lucia	Guatemala Puerto Rico St. Vincent and the Grenadines
America, South			Ecuador	Argentina Colombia	Chile Uruguay
Asia	Myanmar		Afghanistan	China Lebanon	Georgia
Europe		Albania Belgium		Denmark Latvia Romania	Italy Lithuania Slovenia
Oceania	Samoa			American Samoa Guam	Australia

7.2 Classification variables and classes used for tabulations

The WCA 2000 Programme proposed the following priority characteristics to be used as classification variables against which other characteristics should be tabulated:

- Total area of holding;
- Total area of agricultural land;
- Number of livestock;
- Purpose of production;
- Number of permanent workers;
- Land tenure;
- Holder's legal status;
- Size of holder's household;
- Holder's age;
- Holder's sex;
- Irrigation.

All of these 11 variables, except 'Number of permanent workers' and 'Size of holder's household', were essential data items as recommended by the Programme.

For each of the above classification variables, the WCA 2000 Programme recommended certain classes. Country practices on the use of these classification variables, and the classes used, are reviewed in comparison with the recommendations provided by the Programme.

1. Total area of holding

This classification variable envisaged to classify holdings into mutually exclusive classes according to the total area of the holding. The following size classes were recommended by the Programme:

(a) Holdings without land; (b) holdings with land (ha): <0.1, 0.1 - <0.2, 0.2 - <0.5, 0.5 - <1, 1 - <2, 2 - <3, 3 - <4, 4 - <5, 5 - <10, 10 - <20, 20 - <50, 50 - <100, 100 - <200, 200 - <500, 500 - <1 000, 1 000 - <2 500, 2 500 and over.

Most countries used this classification variable in their tabulation Programme. However, none of them followed exactly the recommended classes. Moreover, in practically all cases it was impossible to derive the proposed classes by data aggregation. Most countries did not single out holdings without land, rather including them into the smallest size class. Some countries used their own traditional area units ('feddan' in Egypt, 'acres' in USA and many Caribbean countries, 'rai' in Thailand, 'tomna' in Malta, etc.), which meant their size classes were not comparable with those recommended by the Programme. Box 7.1 presents some country examples of size classes for the classification variable 'total area'.

Box 7.1 Size classes of total area of holding used by some countries

Cote d'Ivoire 2002
<0.5, 0.5-<1, 1-<3, 3-<5, 5-<10, 10-<20, 20-<50, 50+ (ha)

Egypt 1999-2000
0, >0-<1, 1-<2, 2-<3, 3-<4, 4-<5, 5-<7, 7-<10, 10-<15, 15-<20, 20-<30, 30-<100,100+ (feddans, 1 feddan=0.42 ha)

Panama 2001
0, >0-<0.1, 0.1-<0.5, 0.5-<1, 1+ (ha)

Puerto Rico 2002
<10, 10-<20, 20-<50, 50-<100, 100-<175, 175-<260, 260+ (cuerdas, 1 cuerda = appr. 0.4 ha)

Argentina 2002
<5, 5-<10, 10-<25, 25-<50, 50-<100, 100-<200, 200-<500, 500-<1 000, 1 000-<2 500, 2 500-<5 000, 5 000-<7 500. 7 500-<10 000, 10 000-<20 000, 20 000+ (ha)

Ecuador 1999-2000
<1, 1-<2, 2-<3, 3-<5, 5-<10, 10-<20, 20-<50, 50-<100, 100-<200, 200+ (ha)

Georgia 2004-05
0, >0-<0.06, 0.06-<0.1, 0.1-<0.2, 0.2-<0.5, 0.5-<1, 1-<2, 2-<3, 3-<4, 4-<5, 5-<10, 10-<20, 20-<50, 50-<100, 100-<200, 200-<500, 500+ (ha)

Thailand 2003
<2, 2-<6, 6-<10, 10-<20, 20-<40, 40-<60, 60-<140, 140+ (rais, 1 rai=0.16 ha)

Malta 2001
<1, 1-<2, 2-<3, 3-<4, 4-<5, 5-<6, 6-<7, 7-<8, 8-<9, 9+ (tomnas, 1 tomna=0.112 ha)

Romania 2001-02
<0.1, 0.1-<0.3, 0.3-<0.5, 0.5-<1, 1-<2, 2-<5, 5-<10, 10-<20, 20-<30, 30-<50, 50-<100, 100+ (ha)

Australia 2001
<50, 50-<100, 100-<500, 500-<1 000, 1 000-<2 500, 2 500-<25 000, 25 000-<100 000, 100 000-<200 000, 200 000-<500 000, 500 000+ (ha)

Samoa 1999
<1, 1-<2, 2-<5, 5-<10, 10-<20, 20-<50, 50-<100, 100+ (acres, 1 acre= appr. 0.4 ha)

2. Total area of agricultural land

This classification variable, along with cropland and arable land, was proposed by the Programme to offset the limitation of the 'total area of holding' in that the latter classification variable includes some areas not used for agricultural production. Cropland is a useful size criterion for studying holdings engaged mainly in crop production, while total area of agricultural land, including also land under meadows and pastures, is a suitable measure for a holding engaged in crop production and raising livestock. Arable land can be used for classifying holdings engaged mainly in growing temporary crops. Most of countries used either one of these variables or another land use characteristic (cultivated area, utilized agricultural area, productive land area, area of land under permanent crops, sown area, harvested area) as a classification variable for tabulation purposes. Some countries used those variables along with the total area of holding, while some of them restricted their tabulations to specific land use variables.

For the total area of agricultural land the WCA 2000 Programme recommended the same classes as for the total area of holding, while for cropland and arable land the following size classes were proposed:

(a) Holdings without cropland/arable land; (b) holdings with cropland/arable land (ha): <0.1, 0.1 - <0.2, 0.1 - <0.2, 0.2 - <0.5, 0.5 - <1, 1 - <2, 2 - <3, 3 - <4, 4 - <5, 5 - <10, 10 - <20, 20 - <50, 50 - <100, 100 - <200, 200 - <500, 500 - <1 000, 1 000 and over.

The above comments concerning the total area of holding are likewise valid for total area of agricultural land and other land use classifications. Box 7.2 presents some examples of country practices in this respect.

Box 7.2 Size classes of total area of agricultural land or its proxies used by some countries

Guinea 2001
Cultivated area: <0.1, 0.1-<0.2, 0.2-<0.3, 0.3-<0.5, 0.5-<0.75, 0.75-<1, 1-<1.5, 1.5-<2, 2-<3, 3-<5, 5+ (ha)

Réunion 2000
Utilized agricultural area: <1,1- <2, 2-<5, 5-<10, 10-<20, 20+ (ha)

Saint Lucia 1996
Cropland: 0,>0 -<0.5, 0.5-<1, 1-<2, 2-<3, 3-<5, 5-<10, 10-<25, 25-<50, 50-<100, 100+ (acres; 1 acre = appr. 0.4 ha)

United States of America 2002
Harvested area: <1, 1-<10, 10-<20, 20-<30, 30-<50, 50-<100, 100-<200, 200-<500, 500-<1000, 1 000-<2 000, 2 000+ (acres, 1 acre = appr. 0.4 ha)

French Guiana 2000-01
Utilized agricultural area: <1,1-<2, 2-<5, 5-<10, 10-<20, 20+ (ha)

Georgia 2004-05
Total area of agricultural land: 0, >0-<0.06, 0.06-<0.1, 0.1-<0.2, 0.2-<0.5, 0.5-<1, 1-<2, 2-<3, 3-<4, 4-<5, 5-<10, 10-<20, 20-<50, 50-<100, 100-<200, 200-<500, 500+ (ha)

Kyrgyzstan 2002-03
Area of arable land, sown area: 0, >0-<0.05, 0.05-<0.1, 0.1-<0.2, 0.2-<0.5, 0.5-<1, 1-<5, 5-<10, 10-<15, 15-<20, 20-<50, 50-<100, 100-<200, 200-<500, 500-<1 000, 1 000-<2 000, 2 000-<2 500, 2 500+ (ha)
Area under permanent crops: 0, >0-<0.05, 0.05-<0.1, 0.1-<0.2, 0.2-<0.5, 0.5-<1, 1-<2, 2-<3, 3-<4, 4-<5, 5-<10, 10-<20, 20-<50, 50-<100, 100-<200, 200+ (ha)

Denmark 2000
Cultivated land area: 0, >0-<0.1, 0.1-<5, 5-<10, 10-<15, 15-<20, 20-<25, 25-<30, 30-<40, 40-<50, 50-<60, 60-<75, 75-<100, 100-<125, 125-<150,150-<175, 175-<200, 200-<250, 250-<300, 300-<400, 400+ (ha)

Hungary 2000
Productive land area: <0.15, 0.15-<0.5, 0.5-<1, 1-<5, 5-<10, 10-<50, 50-<300, 300+ (ha)

New Caledonia 2002
Utilized agricultural area: <1, 1-<2, 2-<5, 5-<10, 10-<20, 20-<50, 50-<100, 100-<500, 500+ (ha)

3. Number of livestock

This classification variable was proposed by the WCA 2000 Programme as an appropriate size criterion for holdings engaged mainly in raising livestock. The Programme stressed that this classification criterion should be used only for the predominant kind of livestock in the country.

The Programme recommended the following size classes for the various kinds of livestock:

For **cattle/buffaloes:**

a) Holdings with no cattle/buffaloes; (b) holdings with cattle/buffaloes (heads): 1-2, 3-4, 5-9, 10-19, 20-49, 50-99, 100-199, 200-499, 500 and over.

For **sheep/goats/pigs:**

a) Holdings with no sheep/goats/pigs; (b) holdings with sheep/goats/pigs (heads): 1-4, 5-9, 10-19, 20-49, 50-99, 100-199, 200-499, 500 and over.

For **chickens:**

a) Holdings with no chickens; (b) holdings with chickens (heads): 1-9, 10-49, 50-199, 200-999, 1 000-4 999, 5 000-9 999, 10 000 and over.

Most countries tabulated their census results using size classes based on the number of livestock/poultry as classification variables. A number of countries followed the classes recommended by the Programme or at least ensured that the required classifications could be derived by aggregating classes. Only a few countries classified holdings without livestock separately. Some countries used numbers of special types of livestock (like dairy cows, breeding sows) or aggregates (cattle + buffaloes, sheep + goats, total poultry) for their tabulations. Box 7.3 shows some examples of size classes based on numbers of livestock/poultry used by countries.

4. Purpose of production

The WCA 2000 Programme recommended two classes for this classification variable: a) producing mainly for home consumption, and b) producing mainly for sale.

About ten percent of countries tabulated their census results according to this classification variable. Some of them, like Georgia, Lao People's Democratic Republic and Romania used the classes recommended by the Programme, while others made equivalent classifications like commercial/traditional in Botswana, traditional/modern in Comoros, commercial/non-commercial in American Samoa, etc.

5. Number of permanent workers

The WCA 2000 Programme recommended the following classes for this item: (a) Holdings without permanent agricultural workers; (b) Holdings with permanent agricultural workers: one worker, two workers, three workers, four workers, five workers, six workers and over.

Only a few countries tabulated their census results according to the number of permanent workers, and none of them followed strictly the classes proposed by the Programme.

Box 7.4. shows some country practices of classification by number of permanent workers.

Box 7.3 Size classes of livestock numbers used by some countries

Egypt 1999-2000
Cattle + buffaloes, Sheep + goats: 1-2, 3-4, 5-9,10-19, 20-49, 50-99, 100-199, 200-499, 500+

Réunion 2000
Dairy cows, sheep, goats: 1-2, 3-4, 5-9, 10-19, 20-49, 50+
Pigs: 1-2, 3-9, 10-49, 50-99, 100-399, 400+
Chicken: 1-19, 20-49, 50-99, 100-499, 500-2 999, 3000+

United States of America 2002
Cattle: 1-9, 10-19, 20-49, 50-99, 100-199, 200-499, 500-999, 1 000-2 499, 2 500+
Milk cows: 1-4, 5-9, 10-49, 50-99, 100-199, 200-499, 500+
Pigs: 1-24, 25-49, 50-99, 100-199, 200-499, 500-999, 1 000-1 999, 2 000-4 999, 5 000+

Cyprus 2003-04
Cattle: 1, 2, 3-5, 6-9, 10-14, 15-19, 20-29, 30-49, 50-99, 100-199, 200-299, 300-499, 500+
Sheep, goats: 1-4, 5-9, 10-29, 30-49, 50-69, 70-99, 100-199, 200-299, 300-499, 500-699, 700-999, 1000+
Pigs: 1-2, 3-4, 5-9, 10-19, 20-49, 50-99, 100-199, 200-399, 400-999, 1 000+
Poultry: 1-9, 10-19, 20-29, 30-49, 50-99,100-499, 500-999, 1 000-4 999, 5 000-9 999, 10 000+

Kyrgyzstan 2002-03
Cattle: 0, 1, 2-4, 5-9, 10-19, 20-39, 40-49, 50-99, 100-199, 200-499, 500-999, 1 000+
Sheep, goats: 0, 1-4, 5-9, 10-19, 20-39, 40-49, 50-99, 100-199, 200-499, 500-999,1 000-2 999, 3 000+
Pigs: 0, 1-4, 5-9, 10-19, 20-39, 40-49, 50-99, 100-199, 200-499, 500-999, 1000+
Poultry: 0, 1-9,10-49, 50-99, 100-199, 200-499, 500-999, 1000-1999, 2 000-2 999, 3 000-4 999, 5 000-7 999, 8 000-9 999, 10 000+
Horses: 0, 1, 2-4, 5-9, 10-19, 20-39, 40-49, 50-99, 100-199, 200-499, 500+

Lao Peoples Democratic Republic 2002-03
Cattle, buffaloes, pigs: 1, 2, 3-4, 5-9, 10+

Croatia 2003
Cattle, cows, heifers: 1, 2, 3, 4, 5, 6, 7-10, 11-15, 16-20, 21-30, 31-50, 51-100, 101+
Sheep, goats: 1, 2, 3, 4, 5, 6-10, 11-20, 21-50, 51-100, 101-200, 201-400, 401-1000, 1001+
Pigs: 1, 2, 3, 4, 5, 6-10, 11-20, 21-50, 51-100, 101-200, 201-400, 401-1000, 1001+
Sows, gilts: 1, 2, 3, 4, 5, 6-10, 11-50, 51-100, 101-200, 201-400, 401-1000, 1001+
Poultry: 1-10, 11-20, 21-50, 51-100, 101-500, 501-1000, 1 001-3 000, 3 001-5 000, 5 001-10 000, 10 001-50 000, 50 001-100 000, 100 001+

Estonia 2001
Cattle: 1-2, 3-9, 10-19, 20-29, 30-49 , 50-99, 100-199, 200-299, 300+
Sheep: 1-2, 3-9, 10-19, 20-29, 30-49, 50-99, 100+
Goats: 1-9, 10+
Pigs: 1-2, 3-9, 10-49, 50-99, 100-199, 200-399, 400-999, 1 000-1 999, 2 000+
Horses: 1-2, 3-4, 5-9, 10-19, 20+

Norway 1999
Dairy cows: 1-9,10-14, 15-19, 20+
Sheep: 1-19, 20-49, 50-99, 100+
Breeding sows: 1-9,10-19, 20-29, 30-49, 50+
Laying hens: 1-99,100-999, 1 000-1 999, 2 000-2 999, 3 000+

New Caledonia 2002
Cattle, pigs: 1-10, 11-20, 21-50, 51-100, 101-500, 501+
Goats: 1-10, 11-20, 21-50, 51-100, 101-500, 501+
Poultry: 1-20, 21-50, 51-100, 101-500, 501-1000, 1 001-5 000, 5 001+

Box 7.4 Size classes of number of permanent agricultural workers used by some countries

---

Guinea 2001
1 worker, 2 workers, 3 to 4 workers, 5 to 9 workers,10 to 14 workers, 15 workers and over

Togo 1996-97
1 worker, 2 workers, 3 workers, 4 workers, 5 workers and over

Saint Lucia 1996
Without workers, 1 worker, 2 workers, 3 workers, 4 workers, 5 workers and over

Lao People's Democratic Republic 1999
Without workers, with workers

---

6. Land tenure

The WCA 2000 Programme proposed the following classes for this classification variable:

(a) Holdings without land; (b) Holdings operated under one tenure form: owned or held in owner-like possession, rented from others (for an agreed amount of money and/or produce, for a share of produce, in exchange for services, under other rental arrangements), operated on a squatter basis, operated under tribal or traditional communal tenure forms, operated under other tenure forms; (c) Holdings operated under two or more tenure forms.

About half of reporting countries classified their census results according to land tenure. As indicated in Section 6.5, land tenure types used by countries varied considerably and many of them were different from those proposed by the Programme. Some countries provided only a breakdown of the number of holdings by land tenure while other countries provided only the breakdown of the area. Most counties, however, provided both breakdowns (number of holdings and area).

The main issue encountered in classifying holdings by land tenure was how to treat holdings operated under several land tenure forms. The ways countries dealt with this issue can be grouped into four broad categories: a) classification of all holdings with various land tenure forms into one class without further breakdown (American Samoa, Ecuador, Guam, India, Nepal, Northern Mariana Islands, Puerto Rico, USA and US Virgin Islands); b) classification of holdings into mutually exclusive classes according to combinations of land tenure forms present on the holding (Albania, Chile, Guatemala, Italy and Panama); c) classification of holdings into mutually non-exclusive classes according to land tenure forms present on the holding (Egypt, Lao People's Democratic Republic, Portugal, Romania and Spain); d) classification of parcels (Saint Lucia) or fields (Lesotho), instead of holdings, into mutually exclusive classes (parcels and fields are supposed to be under one tenure form). Norway used an original way of classifying holdings according to land tenure: holdings were classified into mutually exclusive classes according to the percentage of rented area: all area rented, 50% or more rented but not all of area rented, <50% of area rented. Box 7.5 presents some country practices of classification by land tenure.

7. Holder's legal status

The following classes were recommended by the WCA 2000 Programme for holder's legal status:

(a) Holdings operated privately: an individual, a household, two or more individuals of different households or two or more households, a corporation, a cooperative, other; (b) Holdings operated by the Government.

About half of reporting countries classified their census results according to the holder's legal status. Quite often, however, the legal status classes used were not in line with those recommended by the Programme. Box 7.6 gives some examples of the legal status classes used by selected countries.

Box 7.5 Country practices of classification by land tenure

India 2000-01 **(classification of holdings with several tenure forms into one class)**
1. Wholly owned and self-operated holdings; 2. Wholly leased-in holdings; 3. Wholly otherwise operated holdings; 4. Partly owned, partly leased-in and partly otherwise operated holdings.

Panama 2001 **(classification of holdings with several tenure forms into mutually exclusive classes)**
a) under single tenure form: 1. owned with title; 2. occupied without title; 3. Rented; b) under mixed tenure form: 1. owned with title and occupied without title; 2. owned with title and rented; 3. occupied without title and rented, 4. owned with title, occupied without title and rented.

Egypt 1999-2000 **(classification of holdings with several tenure forms into mutually non-exclusive classes)**
1. Completely owned; 2. Completely cash rented; 3. Completely rented by partnership; 4.Completely by other forms; 5. Partially owned; 6. Partially cash rented; 7. Partially rented by partnership; 8.Partially by other forms;

Lesotho 1999-2000 **(classification of fields by land tenure forms)**
1. Owned & operated; 2. Owned & share-cropped; 3. Owned & operated by projects; 4. Not owned but operated; 5. Owned but rented out.

Box 7.6 Legal status classes used by selected countries

Egypt 1999-2000
1. Individuals; 2. Corporations; 3. Cooperatives.

Saint Vincent and the Grenadines 2000
1. Individual or household; 2. Partnership; 3. Cooperative; 4. Corporation; 5. Government; 6. Other.

Argentina 2002
1. Physical persons; 2. De-facto societies; 3.Accidental societies; 4. Limited liability societies; 5. Anonymous societies and joint stock companies; 6.Cooperatives; 7.Non-profit private institutions; 8. Public entities

Myanmar 2003
1. Private (household); 2. Cooperate with other household; 3. Corporation/Cooperative; 4. Government; 5. Other

Lithuania 2003
1. Registered farmer's farms; 2. Family farms; 3. Individual enterprises; 4. Agricultural companies; 5. Partnerships; 6. Join-stock, close joint, investment companies; 7. State and municipal enterprises; 8. Cooperative companies.

Guam 2003
1. Individual; 2. Partnership; 3. Corporation; 4. Other.

8. Size of holder's household

The following classes were recommended by the WCA 2000 Programme for the size of holder's household:

1 person, 2 to 3 persons, 4 to 5 persons, 6 to 9 persons, 10 persons and over.

Some ten percent of reporting countries classified their census results according to the size of the holder's household. In some cases, the size classes followed the recommendations of the Programme. Other countries used different size classes some of which permitted aggregation to the recommended size classes, and some of which did not. In some African countries, where traditionally holdings are operated by several households, the data were broken down by size classes based on the number of household members of all households involved in the holding's operation.

Box 7.7 provides some country practices in this respect.

Box 7.7 Size classes of holder's household used by some countries

---

Egypt 1999-2000
1 person, 2 persons, 3 persons, 4 persons, 5 persons, 6 persons, 7 persons, 8 persons, 9 persons, 10 persons, 11 persons, 12 persons,13 persons and over.

Mali 2004-05 **(breakdown by holding size)**
Less than 5 persons, 5 to 9 persons, 10 to 14 persons, 15 to 19 persons, 20 to 24 persons, 25 to 29 persons, 30 persons and over.

Saint Lucia 1996 **(in line with Programme recommendations)**
1 person, 2 to 3 persons, 4 to 5 persons, 6 to 9 persons, 10 persons and over.

Lao People's Democratic Republic 1999
1 to 2 persons, 3 to 4 persons, 5 to 6 persons, 7 to 8 persons, 9 to 10 persons, 11 persons and over

Hungary 2000
1 person, 2 to 3 persons, 4 to 5 persons, 6 to 10 persons, 11 persons and over

---

9. Holder's age

The following classes were recommended by the WCA 2000 Programme for holder's age:

Under 25 years, 25 to 34 years, 35 to 44 years, 45 to 54 years, 55 to 64 years, 65 years and over.

About half of reporting countries classified their census results according to holder's age. Countries chose age classes according to their needs. Box 7.8 shows some country practices.

Box 7.8 Holder's age classes used by some countries

Madagascar 2004-05
15 to 19 years, 20 to 24 years, 25 to 29 years, 30 to 34 years, 35 to 39 years, 40 to 44 years, 45 to 49 years, 50 to 54 years, 55 to 59 years, 60 to 65 years, 65 years and over.

Guadeloupe 2000-01
Under 30 years, 30 to 39 years, 40 to 49 years, 50 to 59 years, 60 years and over.

Chile 1996-97
Under 25 years, 25 to 34 years, 35 to 44 years, 45 to 54 years, 55 to 64 years, 65 to 74 years, 75 years and over.

Thailand 2003
15 to 19 years, 20 to 24 years, 25 to 34 years, 35 to 44 years, 45 to 54 years, 55 to 64 years, 65 to 69 years, 70 years and over.

Norway 1999
Under 29 years, 30 to 39 years, 40 to 49 years, 50 to 59 years, 60 to 69 years, 70 years and over.

Northern Mariana Islands 2003
Under 35 years, 35 to 44 years, 45 to 54 years, 55 to 64 years, 65 years and over.

---

10. Holder's sex

About half of reporting countries have classified their census results according to holder's sex.

11. Irrigation

The following classes were recommended by the WCA 2000 Programme for classification according to irrigation:

Holdings that do not irrigate any land; Holdings that irrigate some land

Few reporting countries classified their census results according to irrigation, and those that did generally went beyond the recommended classification. Box 7.9 presents some examples.

Box 7.9 Country practices of classification according to irrigation

Egypt 1999-2000

Classification according to source of irrigation water. Classes used: Nile water, ground water, agricultural drainage water, rain water, other sources

United States of America 2002

Classification according to size of irrigated land. Size classes used: 1 to 9 acres, 10 to 49 acres, 50 to 99 acres, 100 to 199 acres, 200 to 499 acres, 500 to 999 acres, 1000 to 1999 acres, 2000 acres and more (1 acre= appr. 0.4 ha)

Chile 1996-97

Classification according to irrigation type. Classes used: gravitational irrigation, mechanical irrigation (sprinklers etc.), localized irrigation (drip irrigation etc.)

Cyprus 2003-04

Classification according to irrigation type. Classes used: surface irrigation, sprinklers, droplet irrigation Classification according to size of irrigated land. Size classes used: up to 0.49 ha, 0.5 to -0.99 ha, 1 to 1.99 ha, 2 to 2.99 ha, 3 to 3.99 ha, 4 to 4.99 ha, 5 ha and more

Romania 2001-02

Classification according to area equipped for irrigation and irrigated area. Size classes used: up to 0.1 ha, 0.1 to 0.3 ha, 0.3 to 0.5 ha, 0.5 to 1 ha, 1 to 2 ha, 2 to 5 ha, 5 to 10 ha, 10 to 20 ha, 20 to 30 ha, 30 to 50ha, 50 to 100 ha, 100 ha and more

Classification according to crops irrigated.

Classification according to source of irrigation water. Classes used: ground water, surface water from the holding territory, surface water from out of the holding territory, water supply network, desalinated water, reused water.

Classification according to irrigation method: Classes used: sprinklers, furrows, flooding, others.

Guam 2003

Classification according to source of irrigation water. Classes used: river or stream, lake or private pond, canal or irrigation district public utility, other.

CHAPTER 8
Quality assurance of census data

No matter how well an agricultural census is organized, it is impossible to avoid errors completely. There are two main types of errors: sampling errors and non-sampling errors.

Sampling errors occur when sampling is used because of the fact that only a part of the population is surveyed. However, sampling errors may be estimated and controlled in the sense that they may be reduced by increasing the sample size or improving sample design. When an agricultural census is conducted on a sample basis, it is recommended practice for the implementing agency to indicate, in the census report, the sampling errors for all published estimates.

Non-sampling errors occur in all censuses and surveys. Ideally, they would not be present but in practice they always occur as a result of mistakes committed at various phases of the census work. As noted in the WCA 2000 Programme, 'Non-sampling errors may arise from numerous sources. The census frame or list of holdings may be incomplete or inaccurate; the wording of questions ambiguous or misleading; enumerators may introduce their own biases; respondents may not really know the true answer, or cannot recall the data requested, and others may consciously answer incorrectly; field work may be inadequately organized or supervised; enumerators may lack specific training or unsatisfactory standards may have been used for their selection; completed questionnaires may be lost' (FAO, 1995 p. 15). Efforts should be made to reduce errors arising from these and many other sources. Moreover, the census organizers should be aware of the quality of the data before they are released for public use and should inform data users about data limitations in order to avoid mistakes in decision making.

The present chapter deals with country practices in reducing non-sampling errors. Different control procedures are performed during an agricultural census. The strict control of preparatory census activities, the pre-test of census questionnaires and the taking of a pilot census are all important to avoid errors in questionnaires, manuals, field operations, management of census materials and control procedures. Editing and processing can ensure a 'clean' database for processing and tabulation. But, while much effort is made in a census to obtain and capture accurate data, it is impossible to avoid errors completely.

All countries participating in the WCA 2000 round practised some type of quality assurance procedures and quality checks, and reported on them. The procedures for assuring the quality of census data can be classified in the following way, according to their timing in the census process:

- **Quality assurance during the preparatory phase of the census**: quality check of census frames, pre-test survey, pilot census;
- **Quality assurance during the field work:** checks on the internal consistency of data in all filled census questionnaires (supervision); re-visits to a sample of holdings to verify the information provided by the respondent;
- **Quality assurance after the census enumeration**: consistency checks during data entry, adjustments for non-response and under-coverage, database cleaning, post enumeration surveys (PES), comparisons with other data sources.

The following sections provide examples of country practices in each of these three types of quality assurance procedures with examples shown in boxes.

8.1 Quality assurance during the preparatory phase of the census

All countries made significant efforts to ensure that the census frame accurately covered the target population without omission, duplication or redundancies. Country practices in this respect are reviewed in Chapter 4. However, no matter how much effort is made, it is practically impossible to ensure a 'perfect' census frame. Box 8.1 describes Puerto Rico's experience of accounting for the under-coverage of the list frame by using a sampling procedure from the area frame.

Pre-test surveys and pilot censuses are powerful procedures undertaken at the preparatory stage of the census for reducing non-sampling errors. While the pre-test surveys are mostly confined to the concepts and definitions used and the census instruments (census questionnaires, instruction manuals, etc.), a pilot census is a 'dry run' of the whole census procedures, on a limited scale, and aims to evaluate all aspects of the census operation. A well conducted pilot census provides critical technical input for improved planning of the main census.

Most countries conducted pre-testing of the census instruments but a smaller proportion conducted a pilot census.

Box 8.1 Example of good practice in quality assurance

Puerto-Rico 2002 – **Accounting for the list frame under-coverage**
The 2002 Puerto Rico census of agriculture was conducted by mailing report forms to farm operators on the Census Mail List (CML). However, because of the dynamic nature of mail lists, some farm operators might not be represented in the CML. To account for this under-coverage, an area frame consisting of the entire island of Puerto Rico was sampled. The area frame was stratified according to the intensity of agricultural land use. Primary sampling units (PSUs) were created based on specific size requirements and permanent boundaries. To further enhance sampling efficiency, 'municipios' with similar agriculture were grouped into nine clusters. Within each stratum and cluster, a random sample of PSUs was selected and further sub-divided into target sampling units called segments. Out of approximately 7 500 segments available for sampling, 300 segments were selected.

Prior to actual census data collection, the enumerators, using aerial photos and 'municipio' maps, identified all farm operators within each assigned area segment. They were checked against the CML. If no match was found in the CML for a farm operator, then a census report form was sent to him/her. For those farm operators found in both area segment and the CML, only the CML report was used to avoid duplication. During the pre-screening process, a total of 558 NML (not-in-CML) farm operators were found in the 300 sampled area segments. Of these, only 350 were determined to be actual farm operators.

The CML data, adjusted for non-response, and the NML data, adjusted for sample expansion factors and design weights, were combined to form a single census estimate.

8.2 Quality assurance during the field work

The WCA 2000 Programme stated that: 'The supervisor's work in overseeing enumerators' work, and assisting them to solve problems encountered, is essential for assuring the quality of the census data. The supervisor's presence and inspection of enumerators' work helps prevent carelessness and facilitates error detection and correction while the enumeration is in progress. The best supervision is achieved by constantly working in the field with enumerators. The supervisor should be present at several initial interviews, to detect deficiencies and take immediate remedial action. Subsequent regular visits should be organized to observe at least one interview and inspect a sample of completed questionnaires for completeness and internal consistency. Omissions must be detected and such holdings visited. Unsatisfactory forms may need to be redone, if necessary, with the supervisor's assistance.'

As seen from the census materials presented by countries, all of them followed the WCA 2000 Programme recommendations. Besides the routine supervisor's work, some countries took additional measures to ensure census data quality during the field work. Box 8.2 presents some of the country examples in this respect.

Box 8.2 Examples of good practice in quality assurance

Saint Vincent and the Grenadines 2000 – **Quality assurance during field work**
A quality Control Team (QCT) was created at the Agricultural Census Unit of the Ministry of Agriculture and Labour. The team travelled throughout the country in coordination with the Regional Extension Supervisors to check the work of the enumerators and supervisors. Supervisors were asked to collect samples of questionnaires from the enumerators and submit them to the QCT for checking. Re-interviews were conducted in randomly selected households to confirm that the households were visited and the information was properly recorded.

Lebanon 1998-99 – **Quality assurance during field work**
Special teams consisting of supervisors and controllers were assigned to complete control questionnaires for a random sample of the enumerated holdings. The sample was drawn in parallel to the advancement of the enumeration process. In total 2.5 percent of the questionnaires were checked.

After the questionnaires were validated by the field team (controllers and supervisors), the ultimate checking was carried out by the central team. The checking procedure involved the cross-inspection of the content of the census questionnaires, the control questionnaires and the village questionnaires as well as comparisons with other available information sources. In some cases it became necessary to contact the holder by telephone to verify the data or to re-visit the holder in order to resolve inconsistencies.

8.3 Quality assurance after the census enumeration

During the WCA 2000 round, countries undertook various types of quality assurance procedures after the census enumeration. These procedures included computer-assisted consistency checking and validation of the data, data editing, imputation of missing data, estimation of under-coverage, introduction of adjustment weights for allowing for non-response, validation with other data sources.

Carrying out sample enumerations as quality checks (the so-called post-enumeration surveys – PES) during or just after the main census was strongly recommended by the WCA 2000 Programme as well as by the publication FAO, 1996a. However, only a limited number of countries carried out such a survey.

Box 8.3 shows some country examples of quality assurance after the census enumeration.

Box 8.3 Examples of good practice in quality assurance

Togo 1996-97 – Checks during data entry
In order to strengthen the questionnaire control process, some checks were incorporated into the data entry software. Specifically:

Exhaustiveness checks: to verify that all survey units were captured.
Plausibility checks: to assure that the entered values were inside acceptable or authorized intervals

Consistency checks: to verify consistency of the main variables of the questionnaire
As a result of these checks, the data entry error rate did not exceed 5 percent.

Canada 2001 – Reduction and estimation of under-coverage
The Census of Agriculture was conducted in conjunction with the Census of Population. The Census of Agriculture questionnaire was dropped off along with a Census of Population questionnaire when someone in the household was a farm operator. In order to reduce and estimate possible under-coverage, two follow-up surveys were conducted in Canada.

The first one, Farm coverage Follow-up Survey (FCFS), was aimed at increasing large farms coverage. It identified all large farms in each province on Statistics Canada's Farm Register that may have been missed by the Census of Agriculture. The operators of these farms were contacted and those that had been missed completed their questionnaires over the telephone.

The second one, Coverage Evaluation Survey (CES), was aimed at estimating the coverage of the Census of Agriculture. It selected a random sample of smaller farm operations from Statistics Canada's Farm Register for which no questionnaire was received. The survey used a short questionnaire to collect key information about the operating status and the size of the farm. The CES estimated the overall under-coverage rate as being 5.6%

Portugal 1999-2000 – Post-enumeration quality survey
A post-enumeration quality survey was conducted aimed at estimating observation errors and assessing the reliability of the census results. The size of the sample used for the quality survey equalled 15 000 which represented 2.36% of the total population of holdings. In order to assure the trustworthiness of the quality survey, its data collection framework was completely independent of that of the agricultural census. The time lag between the census and quality survey interviews did not exceed two weeks.

Any difference between the census and quality survey results was interpreted as an error committed by the census enumerator. For quantitative variables, the results were grouped in value classes and those classes were compared to avoid random and insignificant differences in values. For each main variable a Global Consistency Index (GCI) was calculated indicating the overall coincidence rate of the results. A table showing the GCI of the main variables was included in the census report. As shown by the GCIs, for most variables the quality of the census results was satisfactory.

CHAPTER 9
Dissemination of census results

A census is not completed until the information collected is made available to users in a form suited to their needs. As the WCA 2000 Programme pointed out, 'census taking uses public resources and the published results represent the public return of a major product from this expenditure'. The role of census dissemination is to enable the public to access and use census data in their decision making. The dissemination Programme, including the publication list, is as important as other components of the census operations.

The WCA 2000 Programme emphasized the need for the census results to be published as soon as possible. Publication of a short preliminary report with advanced census data, several volumes for the general census report, and details of organizational and administrative aspects of the census were some of the recommendations made by the WCA 2000 Programme with respect to the publication of census results. Moreover, where sample enumeration was used, the census report should provide all relevant sample design details.

The WCA 2000 Programme recommended that the priority tables should be released as soon as possible, including data from all holdings enumerated, but with limited cross tabulations. Further analysis could be provided later by releasing additional cross tabulations, making available disaggregated data to users for special analyses and making available facilities for the production of special tables requested by users below the level released in the census report. However, measures should also be taken to safeguard data confidentiality and avoid data disaggregation below the level that professional statisticians would accept as valid.

9.1 Time gap between enumeration and publication

As recommended by the WCA 2000 Programme, countries should make every effort to reduce the time gap between the census enumeration and the publication of the census results. This interval can be shortened by publishing preliminary results from manual processing, or the summation of the field control records, with the full and detailed results following at a later date.

Detailed information about the year of publication of census reports is provided in Annex Table A3. Table 9.1, based on that table, summarizes by region the time gap between the end of enumeration and the publication of the main census report. The country practices show that many efforts were made to ensure the immediate availability of census results: about 40% of census data were published in the same year as the enumeration ended. Up to two thirds of censuses published their final report the same year as the enumeration ended or the next year. Only in 14 cases, out of the 114 censuses for which reports were available, did the publication take three years or more.

Table 9.1 Distribution of number of censuses of the WCA 2000 round according to the time gap between the end of field enumeration and the publication of census results

Region	Total number of reported censuses	Same year	Next year	2 years	3 years	More than 3 years	Without information
All reported censuses	114	44	30	24	12	2	2
Africa	25	5	8	7	4	0	1
America, North and Central	14	9	4	1	0	0	0
America, South	8	3	3	1	0	1	0
Asia	29	9	7	8	3	1	1
Europe	29	14	4	6	5	0	0
Oceania	9	4	4	1	0	0	0

9.2 Country practices

Almost all countries which undertook an agricultural census in the WCA 2000 round, issued a publication with their census results. In only eight cases did FAO not have access to the census report. The type and depth of the census publications were, however, very variable ranging from more than six volumes in Senegal to one small volume in many of the Caribbean countries.

Advanced census results were produced by the majority of countries. Details of the census organization, timetables and other administrative aspects as well as census methodology and main definitions and concepts were also common in census reports.

The use of technology has been evolving over previous rounds of censuses, transforming data presentation, storage and communication. During the WCA 2000 round, there was an increased use of micro-computers for data processing and the preparation of reports and tables for dissemination. Many countries took advantage of CD-ROMs to disseminate census results which is a cheaper and less bulky way of dissemination than by means of hard copy publications. Many countries made available special tabulations upon request, some of them for a fee and some without any fee. Some countries put their census reports on the internet thus making them available to a wide range of users, while some of them, as seen from the example in Box 9.1, took advantage of the internet to enable users to create their own queries.

Box 9.1 An example of good practice in reporting census resultss

---

Italy 2001
In accordance with the dissemination plan, the final results of the 5th General Census of Agriculture were published in the following way:

A. General publications
1. Structural characteristics of agricultural holdings
   a. National publication
   b. Regional publications
   c. Provincial publications
2. Typological characteristics of agricultural holdings
   a. National publication
   b. Regional publications
   c. Agricultural censuses in Italy from 1961 to 2000
3. Organization and documents of the 5th General Census of Agriculture

B. Thematic publications
1. Viticulture in Italy
   Volume 1 - General characteristics
   Volume 2 – Vine varieties
2. Family holdings
3. Agricultural enterprises
4. The young in the agriculture
5. Land use (plant growing)
6. Animal husbandry
7. Holding infrastructure (agricultural machinery, equipment and buildings)
8. Women in agriculture

All the above mentioned volumes were made available on the website of Istat (The National Institute for Statistics of Italy)

C. Data Warehouse
After the end of the checking and control process, the data were stored in a specialized data warehouse, that is, they were transformed from the 'questionnaire' to a relational database more suitable for micro-data queries. The data warehouse allowed remote users to create their own queries and obtain answers to them in compliance with the statistical data confidentiality requirements.

CHAPTER 10
Integration of the census of agriculture with other censuses

As an integral part of a national statistical system, the census of agriculture is related to many other statistical operations in a country. In several parts of the WCA 2000 Programme, mention is made of other sources complementing the structural data on agriculture from the census. For example, in paragraph 1.11 the Programme states: 'Continued emphasis is placed on the need for countries to prepare a multi-year plan for a sequential statistical Programme of activities relating to data collection, processing and analysis, and to allocate adequate resources in a balanced manner for the components of the Programme'.

This chapter refers mainly to country experiences in coordinating the agricultural census with other major data collection activities, in particular: population and housing census, forestry and aquaculture censuses, and community-level data collection which may be considered as a community census. This chapter consists of four sections: Section 10.1 refers to the relationship between the agricultural and the population and housing census. Sections 10.2 and 10.3 deal with the linkages between the agricultural census and forestry and aquaculture/fisheries censuses, respectively. Section 10.4 provides examples of country practices of using the agricultural census for collecting community level data.

10.1 Census of agriculture and census of population

Country practices in the WCA 2000 round show several ways of linking agricultural and population and housing censuses. These can be categorized into three groups:

a. The population census does not cover agricultural variables but other information is used to construct a frame for the agricultural census, especially the household lists;
b. The population census collects some information on agricultural variables and serves as a screening procedure for detecting agricultural holdings;
c. The population census and the agricultural census are undertaken together (as a single census operation).

Information from a previously conducted population census is widely used in the preparation of the agricultural census, even if no special provisions were made for including agriculture related questions in the population census questionnaire. Data on the number of rural households by administrative division are crucial when planning an agricultural census in order to define enumeration areas, workloads, logistics, etc. Also, a precise definition of the rural area is very important when planning the agricultural census. If the agricultural census is undertaken on a sample enumeration basis, enumeration areas from the population census are often used as PSUs. Sometimes a list of households from the population census is directly used during the agricultural census for screening households for agricultural holdings. In Annex Table A6 such census frames are marked as 'List of HH from a PC'. China used a list of households as obtained from the 1990 population census as a frame for the agricultural census in 1997. In Jordan, a complete list of all households from the 1994 Population Census, whether or not engaged in agriculture, was used as the frame for the Agricultural Census 1997. The households engaged in agriculture were then identified during the census interviews. For the Agricultural Census 1999 of the Lao People's Democratic Republic, the census frame was prepared using the list of villages and households collected during the 1995 Population Census. In the Agricultural Census 2002 of Sri Lanka, the census frame for the small holding sector was the same list of households as prepared for the 2001 Population Census. All these censuses may be classified in the above mentioned category (a).

Sometimes, as classified in the above category (b), the population census contained some screening questions aimed at identifying agricultural holdings. The identified holdings were afterwards visited either on a complete or a sample enumeration basis. In the questionnaire of the 1998 Population and Housing Census of Cote d'Ivoire appropriate questions were inserted to identify and list agricultural households.

The list was subsequently used for defining the sampling frame for the holdings in the Traditional Sector. For the Libyan Agricultural Census 1999-2000, a list of agricultural holdings and holders was obtained from the 1995 Population Census whose questionnaire included some questions aimed at identifying households engaged in agricultural activities. In Trinidad and Tobago the Visitation Records of the 2000 Population and Housing Census reported some questions specifically designed to identify individuals and households involved in agricultural activities and to capture the basic characteristics of the agricultural holdings. These records were used as one of the sources for the construction of the frame for the 2004 Agricultural Census of Trinidad and Tobago. Box 10.1 shows some examples of questions included in population census questionnaires to identify agricultural holdings.

The strongest link between the two censuses is when both censuses are undertaken together as a joint operation. During the WCA 2000 round this was done either by inserting an agricultural module in the population census questionnaire, as in Serbia, Seychelles, Uganda, and Zambia, or by conducting a full scale agricultural census together with the population census, as in Canada and Poland. Box 10.2 provides some country examples.

Box 10.1 Screening questions included in the Population Census to identify agricultural holdings

---

Botswana 2004
In the 2001 Population and Housing Census (2001 PHC) the following questions were included:

Q1. Does any member of this household own any of the following livestock?
   (1)Cattle (2) Goats (3) Sheep (4) Pigs (5) Poultry (6) Donkeys/horses
Q2. Does any member of this household plant any of the following crops during the agricultural season?
   (1)Maize (2) Millet (3) Sorghum (4) Beans (5) Other crops
Q3. Does this household own the land used for planting and/or grazing? (Yes/No)
Q4. How was the land used for planting and/or grazing acquired?
Q5. Since 1999, did household members receive cash from sale of
   (1)Cattle (2) Goats/sheep (3) Poultry (4) Maize (5) Sorghum/millet?

These questions were used for creating the sampling frame for the Agricultural Census 2004.

Nepal 2002
During the listing operation of the 2001 Population Census the following questions were included

Q1. Is any agricultural land possessed by the household? (Yes/No)
Q2. If yes, what is the area by land type?
Q3. Does your household raise livestock or poultry? (Yes/No)
Q4. If yes, how many livestock/poultry does your household raise?
   (1) Livestock (2) Poultry

The frame of agricultural holdings created with the help of these questions was used in the National Sample Census of Agriculture of Nepal

---

Box 10.2 Population Census and Agricultural Census undertaken as a joint operation

---

Uganda 2002

An agricultural module (AM) was piggy-backed onto the Population and Housing Census of Uganda. The main purpose of the AM was to provide appropriate sampling frames for a planned Uganda Census of Agriculture and Livestock (UCAL) and other surveys. The AM was administered on a universal basis and therefore covered all households engaged in agricultural activities.

The AM collected information about the following items:

- Whether any member of the household is engaged in any of the following activities: crop growing, livestock rearing, poultry keeping and fish farming;
- Total area of the holding;
- Number of plots by crop grown and by pure or mixed stand during the agricultural season of January-June 2002;
- Number of livestock by type as of the enumeration day;
- Average number of poultry reared per month by type in the last three months;
- Number of fish ponds by type of fish stocked and number of un-stocked fish ponds.

Canada 2001

When questionnaires for the Census of Population were distributed to households across Canada, the Census Representative (CR) asked whether anyone in the household was a farm operator. If so, the CR left a Census of Agriculture Form (CAF) for completion. The CAF was left also if no contact was made but the CR saw evidence of agricultural activity. In addition, the Census of Population questionnaire had a separate question asking if anyone was a farm operator. If someone answered 'yes' to the question, the CR either confirmed that a CAF had been delivered to the household or made arrangements to deliver it. The CAF contained 16 pages and was quite comprehensive.

10.2 Census of agriculture and census of forestry

The WCA 2000 Programme recommended that economic units engaged solely in forestry and logging activities should not be considered as agricultural holdings and should be excluded from the census. For the agricultural holdings for which forestry was a secondary or ancillary activity, the Programme recommended a restricted range of items, namely, whether forest trees existed on the holding, total area under forest trees, area reforested during the last 5 years, age of trees, whether forest products were harvested, and value of sales.

For some countries, however, forestry and logging activities were quite important, so they included economic units engaged solely in forestry in the census coverage and collected more detailed information about these activities, so it may be considered that their agricultural census included a small-scale census of forestry. Japan conducted a full-scale forestry census jointly with the agricultural census, officially naming its census activity 'Census of Agriculture and Forestry'. In Estonia a special volume was dedicated to forestry and fisheries tables. Some countries, like Malaysia, Slovenia and Uruguay, included questions about forestry in their census questionnaires but because their holding criteria did not refer to forestry activities (see Annex Table A8), the complete forestry sector was not covered. Box 10.3 presents some examples of the integration of censuses of agriculture and forestry.

Box 10.3 Examples of integration of agricultural and forestry censuses:

Japan 2005

The census was officially named 'The Census of Agriculture and Forestry'.

At the holding level, three questionnaires were filled: 'Questionnaire for forestry households', 'Questionnaire about forestry holdings that are not family-owned', and 'Questionnaire about forestry service establishments'.

Besides, two questionnaires: 'Survey of land area of forests - A' and 'Survey of land area of forests - B' were completed for each municipal body.

Chile 1996-97

A separate section (Seccion V: Plantaciones forestales) was dedicated to forest plantations in the census questionnaire, collecting information about areas of specific forest trees (poplar, acacia, eucalyptus, pine, evergreen beech etc.) as well as the area of forestry nurseries.

The census was conducted on a complete enumeration basis by screening enumeration areas. Minimum size criteria ensured that holdings with at least 0.1 ha of wood plantations were eligible for enumeration, so it can be considered that in Chile the agricultural census contained a small component of a forestry census.

Estonia 2003

The agricultural holding was defined as a production unit with at least one hectare of agricultural or forest land or at least 0.3 hectare of fish ponds. Therefore, economic units engaged solely in forestry were covered by the census provided they were sufficiently large. For each holding with at least 1 hectare of forest land a 'Forestry Questionnaire' was completed. The questionnaire collected information about area of forest by type (young growth, lath forest, middle-aged forest, growing forest, ripe forest, undetermined), total reserve of stands in cubic meters broken down by coniferous and broadleaved trees, forest felling in cubic meters by cutting type (regeneration cutting, improvement cutting, selection cutting, deforestation), reforested area (in cutting areas, in former quarries, in burnt woodland and glades, in arable land and pastures), length of forest roads, drained area, number of machinery and equipment by type (power saw, brush cutter, device for collecting timber, forwarder, harvester).

Special tabulations were made based on the data collected from the Forestry Questionnaire. They made part of a separate volume of census publications: 'Agricultural Census IV. Forestry. Fishing.'

10.3 Census of agriculture and aquaculture and/or fishing census.

The WCA 2000 Programme recommended that economic units engaged solely in aquaculture and/or fishing activities should not be considered as agricultural holdings and should be excluded from the census. For the agricultural holdings with aquaculture and/or fishing as a secondary or ancillary activity, the Programme recommended a restricted range of items, namely, whether these activities existed on the holding, type of aquaculture installation (pond, rice field, other), kind of aquaculture products and value of sales.

However, quite a number of countries for which fisheries activities were important in their own right, not only in connection with agriculture, went beyond this narrow range of items, including fishing and/ or aquaculture modules in their census questionnaires. In most of these cases, the census was conducted by screening enumeration areas or lists of households and no minimum size criteria were imposed on agricultural holdings, so all aquaculture and fishing holdings were covered, at least in the household sector. Therefore, such censuses may be considered as examples of the integration of small-scale aquaculture and/

or fishing census with the agricultural census by inserting an aquaculture and/o fishing module into the agricultural census questionnaire. In some cases, like in the Philippines and Saint Kitts and Nevis, the census activity was even officially called the Census of Agriculture and Fisheries. In Myanmar aquaculture holdings were identified and enumerated.

Box 10.4 shows some examples of the integration of censuses of agriculture and aquaculture/fishing.

Box 10.4 Examples of integration of agricultural and aquaculture/fishing censuses:

Tanzania 2004

A section about fish farming was included into the census questionnaire. The section collected information about the fish farming system (natural pond, dug out pond, natural lake, water reservoir or other), size of unit/pond in square meters, source of fingerlings (own pond, government institution, NGO/Project, neighbour, private trader or other), frequency of stocking (number of times per year the farmer puts new fingerlings into the pond), number of stocked fish, number of fish harvested, weight of fish harvested, weight of fish sold, to whom sold mainly (neighbour, local market, secondary market, processing industry, large-scale farm, trader at farm, did not sell, other).

The census was conducted on two-stage sampling basis with households as secondary sampling units. Therefore the census covered even those units which were engaged solely in aquaculture (fish farming) and thus can be considered as an example of a small-scale sample aquaculture census (at least at household level) conducted jointly with a sample agricultural census.

Guatemala 2003

An entire section (Capitulo IX. Actividad acuícola) of the census questionnaire was dedicated to aquaculture. It collected information about the number of ponds, quantity of production of specific aquaculture products (shrimps, fish, frogs, snails etc.) as well as the surface area of the water body dedicated to each of them. In addition, information was collected about aquaculture practices utilized (use of concentrated feed, density control of aquatic organisms, sanitary control, water quality control, use of unisex fry).

The census was conducted on a complete enumeration basis by screening enumeration areas with no cut-off limits for agricultural holdings. Therefore the census can be considered as an example of a small scale complete enumeration aquaculture census (at least at the household level) conducted jointly with an agricultural census.

Myanmar 2003

Although the census was officially named 'Myanmar Census of Agriculture 2003' (MCA 2003), Form 2003 MAC-4 – 'Household Fishing Questionnaire' and Form 2003 MAC-5 – 'Aquaculture Holding Questionnaire' were specially developed and included in the census tools. This was done because of the importance of fish and fishery products in the diet of the Myanmar people.

The Household Fishing Questionnaire recorded information on fishing location (paddy field, river, canal stream, lake/pond, mangrove creek, sea/close to shore, other), age group (under or over 10) and sex of the household members engaged in fishing, type of gear used (hand/no gear, cast net, rod, small trap, lift net/ push net, gill net, large fixed trap, other), main purpose of production (for sale, for home), number and type (powered, not powered) of boats used (if used at all).

The Aquaculture Holding Questionnaire recorded information on the aquaculture holding, specifically: holding location and holder's demographic data, total area of production units, type of water used (fresh, brackish, seawater), principal species cultured and whether they were stocked with fingerlings, fed, and sold, machinery/equipment used, whether there were agricultural crops around the aquaculture holding.

10.4 Census of agriculture and community level data.

The collection of community level data was not included among the recommendations of the WCA 2000 Programme. However, some countries developed a questionnaire for collecting data at community (village, commune) level and added it to other census questionnaires aimed at data collection at the holding level. These countries took advantage of the fact that the community administration was involved in the census operations (the listing process of households/holdings or data collection itself), so in these circumstances community level data could be collected at little additional cost.

Community level data are of statistical interest for three main reasons: a) they are of interest in their own right; b) they can be useful for analysis in relation to holding level data tabulated at community level; c) the data may be used for checking holding level data collected during the agricultural census.

Box 10.5 below describes some country practices of the collection of community level data.

Box 10.5 Examples of community-level data collection:

Madagascar 2004-05

Commune Profile Questionnaire (Questionnaire Monographie des communes) was included in the census tools. The questionnaire collected information about:

- Villages (fokontany) in the commune and their population
- Land use
- Economic activities of the Commune (agriculture, trade, transportation, industry and handicrafts, etc.)
- Social and educative life
- Infrastructure

Lebanon 1998-99

To comprehend all aspects connected with agriculture, a village questionnaire was developed. The questionnaire covered the following items:

- Area of permanently fallow land, forests and uncultivated land
- Potential development of uncultivated lands and forests
- General information about the village inventory of problems and difficulties encountered at the level of infrastructure, institutions, economic activities and natural resources
- Main projects envisaged
- Members of municipal councils and/or main decision makers were interviewed in each village to fill the questionnaire, in parallel to the census field work.

Japan 2005

A questionnaire named 'Survey on Rural Communities' was part of the census tools and included the following groups of items:

- Geographical conditions
- Number of houses
- Cultivated lands
- Agricultural production
- Practices (agricultural associations, community gathering, management of facilities)
- Preservation of regional environment resources
- Life environment

PART TWO

Annex
Detailed tables

Notes to the Annex Tables

Notes to Table A2
(1) The year of participation refers to that in which the census enumeration took place. If the enumeration took place over several years, all are shown separated by a dash. If a country has conducted more than one census during a round, all are shown separated by '&' (for some European countries taking agricultural censuses annually, the year of the round is indicated with the word 'annual' in square brackets next to it).
(2) For comparability purposes, in the calculation of the sub-totals and totals the censuses conducted by special estimates are not taken into account

Notes to Table A3
(1) In case a country has conducted more than one census during the 2000 round, all columns of the table except 'Time gap' refer to the census closest to the middle of the round reported in the publication FAO, 2010.
(2) In calculation of time gaps the following simplifying assumptions were made: (a) if a country has conducted several censuses during the 2000 round, the time gap indicated is between the last two censuses of the round; (b) if a census was taken over more than one year, the last year of the census (when it was actually completed) is used in the calculations; (c) censuses conducted by special estimates were ignored; (d) Croatia, Czech Republic, Serbia and Slovakia took their first censuses as independent countries during the 2000 round; however, all their territory, as part of a former country (Czechoslovakia and Yugoslavia), was covered by a census in a previous round, the so time gap indicated is between those censuses; (e) Yemen conducted its first census as a united country in 2003 but all its territory was covered by censuses conducted in 1983 and 1984 in Yemen Arab republic and Yemen People's Democratic Republic, respectively, so the time gap for Yemen was taken as 19 years.

Note to Table A6
(1) For explanation of texts in the cells, see Section 4.2 of Chapter 4.

Notes to Table A8
(1) The columns show groups of variables used for establishing thresholds. Specific variables used are indicated in the cells.
(2) Only the censuses that have reported a threshold are listed. For each census, minimum thresholds for inclusion are indicated in the corresponding row. To compile the overall inclusion criterion, the criteria defined in the row must be connected with the conjunction 'or'. For instance, the inclusion criterion for Tunisia will read as follows – a holding was covered by the census if it had at least: 0.01 ha under irrigated cultures; or 1 ha under non-irrigated cultures; or one cow; or ten female camels; or six female sheep and goats in total; or ten mares; or 5 000 chickens and 50 laying hens; or 50 adult female-rabbits, five cages of adult males and five cages of young rabbits.

Abbreviations Used in the Annex Tables A4 and A6

AH - Agricultural Holding; D - District (or analogous administrative unit); EA - Enumeration Area; ep - selection with equal probabilities; HH - Household; NA (na) - information not available; O – other type of sampling unit; PC - Population Census; PFU - Primary Frame Unit of an area frame; pps – selection with probabilities proportional to size; p1 - selection with certainty (probability 1); S - Segment of an area frame; V - Village (or analogous administrative unit).

Table A1 - Comparison of items between the 2000 and 1990 WCA Programmes

Items as in the WCA 2000 Programme	Item type in the WCA 2000 Programme	Item type in the WCA 1990 Programme	REMARKS
Category 01: Identification			
011 Holding - Address	Essential	Essential	In the WCA 1990 Programme in this category there was an additional essential item 'Head of household's name'

012 Holder
  - Name
  - Address, if different from holding
	Essential
	Recommended
		Essential
		Recommended	

013 Respondent for holding
  - Whether respondent is holder
  - Name, if not holder
	Recommended
	Recommended
		Recommended
		Recommended	

Category 02: General Characteristics			
021 Holder - Legal status of holder	Essential	Essential	

022 Hired Manager
  - Whether there is a hired manager
  - Name
  - Address, if different from holding
  - Economic status of hired manager
	Recommended
	Recommended
	Recommended
	Recommended
		Recommended
		Recommended
		Recommended
		Recommended	

023 Economic activities of an enterprise
  - Whether holding is part of enterprise engaged also in other economics activities
  - Other economic activities of enterprise
	Recommended
	Recommended
		Recommended
		Recommended
			In the WCA 2000 Programme this group of items is marked as having 'environmental implications'

024 Purpose of production	Essential	Not in the list of items	
Category 03: Demographic Characteristics			
031 Household - Number of household members	Recommended	Essential	In the WCA 1990 this category is called 'Demographic and Anthropometric Characteristics'

032 Household members
  - Name
  - Age
  - Sex
  - Marital status
  - Education
	Recommended
	Essential
	Essential
	Recommended
	Recommended
		Recommended
		Essential
		Essential
		Not in the list of items
		Not in the list of items
			In the 1990 WCA the essential character of age and sex referred only to the holder

Category 04: Employment			

041 For each household member
  - Whether economically active or not
  - Main occupation
  - Whether engaged in more than one occupation
  - Whether any work done on olding during the year
  - Whether permanent or occasional agricultural worker in the holding
	Recommended
	Essential
	Essential
	Recommended
	Recommended
		Recommended
		Essential
		Essential
		Recommended
		Recommended
			In the 1990 WCA the essential character of the items in this category referred only to the holder

042 Agricultural workers other than members of holder's household
  - Whether permanent/occasional agricultural workers employed during year
  - Number of permanent agricultural workers male/female
	Recommended
	Recommended
		Recommended
		Recommended	

Table A1 - Comparison of items between the 2000 and 1990 WCA Programmes

Items as in the WCA 2000 Programme	Item type in the WCA 2000 Programme	Item type in the WCA 1990 Programme	REMARKS
Category 05: Land and Water			
051 For entire holding			
- Number of parcels	Essential	Essential	
- Total area	Essential	Essential	
- Whether land was rented to others	Recommended	Recommended	
- Area of land rented to others	Recommended	Not in the list of items	
052 For each parcel			
- Location	Recommended	Recommended	
- Total area	Essential	Recommended	
- Land tenure	Essential	Essential	
- Land use	Essential	Essential	
- Whether irrigated at some time during the year	Essential	Essential	In the 1990 WCA for the entire holding
- Whether drainage facilities available	Recommended	Not in the list of items	In the 1990 WCA for the entire holding
- Area irrigated	Recommended	Not in the list of items	In the 2000 WCA these two items are marked as
- Area affected by salt soil or high water table	Recommended	Not in the list of items	having 'environmental implications'
- Area with irrigation potential	Recommended	Not in the list of items	
- Whether shifting cultivation practised	Essential	Recommended	In the 1990 WCA for the entire holding
- Year current parcel cleared for cultivation	Recommended	Recommended	
- Soil (type, colour, depth, salinity, surface drainage, rate of percolation, degradation, relative area of degradation)	Recommended	Not in the list of items	In the 2000 WCA this group of items is marked as having 'environmental implications'
Category 06: Crops			
061 Temporary crops			
- Name of crops grown	Essential	Recommended	In the WCA 1990 some crop characteristics are
- Area harvested	Essential	Essential	recommended to be collected also for each parcel
062 Permanent crops			
- Name of crops grown	Recommended	Recommended	
- Number of scattered trees	Recommended	Recommended	
- Number of trees of productive age in compact plantations	Essential	Recommended	
- Area under trees of productive age in compact plantations	Essential	Recommended	
- Area under trees of non-productive age in compact plantations	Recommended	Recommended	
063 Fertilizers			
- Whether inorganic fertilizers applied	Essential	Essential	
- Whether organic manure or other fertilizers applied	Recommended	Recommended	
- Amount of inorganic fertilizers applied per crop	Recommended	Not in the list of items	In the 2000 WCA this item is marked as having 'environmental implications'
064 Pesticides			
- Whether applied during year	Essential	Recommended	
- Frequency of pesticide applications per crop	Recommended	Not in the list of items	In the 2000 WCA this item is marked as having 'environmental implications'
065 Seeds and young plants			
- Whether high yield variety seeds used during year	Essential	Recommended	
- Crops with high yielding varieties of seeds	Recommended	Not in the list of items	In the 2000 WCA this item is marked as having
- Crops with traditional varieties of seeds	Recommended	Not in the list of items	'environmental implications'
Category 07: Livestock			
071 Livestock production system			
- Type of livestock production system	Essential	Recommended	

Table A1 - Comparison of items between the 2000 and 1990 WCA Programmes

Items as in the WCA 2000 Programme	Item type in the WCA 2000 Programme	Item type in the WCA 1990 Programme	REMARKS
072 Animals numbers, by type, age, sex and purpose (for each relevant kind of livestock in the country)			In the 1990 WCA only total number of animals was included as essential item. No recommendations were given about breakdown by age, sex and purpose
- Cattle	Essential	Essential	
- Buffaloes	Essential	Essential	
- Sheep	Essential	Essential	
- Goats	Essential	Essential	
- Pigs	Essential	Essential	
- Horses	Essential	Essential	
- Camels	Essential	Essential	
- Mules and hinnies	Essential	Essential	
- Asses	Essential	Essential	
073 Poultry			
- Chickens	Essential	Essential	
- Ducks	Recommended	Not in the list of items	
- Geese	Recommended	Not in the list of items	
- Turkeys	Recommended	Not in the list of items	
- Guinea fowls	Recommended	Not in the list of items	
- Pigeons	Recommended	Not in the list of items	
- Other poultry	Recommended	Recommended	
074 Other domesticated animals			
- Beehives and bee colonies	Recommended	Essential	
- Rabbits and hares	Recommended	Recommended	
- Lamas and alpacas	Recommended	Recommended	
- Fur-bearing animals	Recommended	Recommended	
- Other n.e.s.	Recommended	Recommended	
Category 08: Machinery and Equipment			
081 Stationary power production machinery			In the 1990 WCA collecting the number used was recommended for all types of machinery
- Number used on the holdings by type	Recommended	Recommended	
082 All other machinery and equipment			
- Whether used on holding by main source and by type	Recommended	Recommended	
Category 09: Building and other structures			
091 Non residential building			
- Whether any non-residential buildings used	Recommended	Recommended	
- Tenure(for each non-residential building used)	Recommended	Recommended	
- Area or volume	Recommended	Recommended	
Category 10: Other activities			
101 Forestry			In the 1990 WCA total number of forest trees was a recommended item
- Existence of forest trees on holding	Recommended	Recommended	
- Area under forest trees	Recommended	Not in the list of items	In the 2000 WCA these two items are marked as having 'environmental implications'
- Age of trees	Recommended	Not in the list of items	
- Area reforested last 5 years	Recommended	Not in the list of items	
- Whether forest products are harvested or not	Recommended	Not in the list of items	In the 2000 WCA this item is marked as having 'environmental implications'
- Value of sales	Recommended	Not in the list of items	

Table A1 - Comparison of items between the 2000 and 1990 WCA Programmes

Items as in the WCA 2000 Programme	Item type in the WCA 2000 Programme	Item type in the WCA 1990 Programme	REMARKS
102 Fisheries - Whether fish or other aquatic animals and plants are taken from the waters within the holding	Recommended	Essential	
- Indication of type o aquaculture installation used for fisheries	Essential	Essential	
- Kind of products	Essential	Not in the list of items	
- Value of sales	Essential	Not in the list of items	In the 2000 WCA this item is marked as having 'environmental implications'
103 Other activities	Recommended	Not in the list of items	

Table A2 - Agricultural censuses conducted in the WCA 2000 and previous WCA rounds.

Countries by Region	Year(s) of participation in the WCA round of						
	2000	1990	1980	1970	1960	1950	1930
AFRICA							
Algeria	2001			1973		1950-51	1930
Angola					1961		
Benin		1992-93					
Botswana	2004	1993-94	1982	1969	1962[1]	1950[1]	
Burkina Faso		1993			1961[2]		
Burundi							
Cameroon			1985	1972			
Cape Verde	2004	1988	1981				
Central African Republic			1985	1973	1960		
Chad				1972			
Comoros	2004						
Congo, Democratic Republic of the		1988-89		1971[3]		1950[4]	
Congo		1986[5]	1977[5]	1972[5]	1960[6]		
Côte d'Ivoire	2002			1973-74			
Djibouti		1995					
Egypt	1999-2000	1990	1982	1960-1961[7]		1950	1929
Equatorial Guinea							
Eritrea[8]							
Ethiopia	2001-02	1988-89	1977				
French West Africa[9]							1929-30*
Gabon				1973-74	1960		
Gambia	2002					1950*	
Ghana			1984	1970	1964	1950[10]*	
Guinea	2001	1989 & 1995		1974-75	1964		
Guinea-Bissau		1988-89			1960-61[11]	1953[11]	
Kenya			1979	1969-70	1961	1954	1930
Lesotho	1999-2000	1989-90	1980	1970	1960[12]		
Liberia				1971			
Libya	1999-2000	1987		1974	1960[13]		
Madagascar	2004-05		1985		1961-62		
Malawi		1993	1981	1969	1960-61[14]	1950[15]	1929-30[15]*
Mali	2004-05		1984		1961		
Mauritania			1985				
Mauritius						1950	1930
Morocco	1996			1974	1962		
Mozambique	2000-01					1951	1930
Namibia	1996-97	1995			1959-60[16]		
Niger			1980		1960		
Nigeria				1974-75		1950[17]	1929-30*
Réunion	2000	1988-89	1981	1973			
Rwanda			1984				
Saint Helena						1950*	
Sao Tome and Principe		1990					
Senegal	1998-99				1960		
Seychelles	1998				1960	1950	1929-30

* Special estimates. The censuses of those countries and territories that during the 1930 and 1950 rounds (and Fiji in the 1960 round) could not take a proper census, but made other efforts to obtain the required minimum information, were also considered as participants. These censuses are not taken into account when counting the totals.
1 Then called Bechuanaland
2 Then called Upper Volta
3 Then called Zaire
4 Then called Belgian Congo
5 Then called Congo
6 Then called Congo Brazzaville)
7 Then called United Arab Republic
8 Independent state since 1993; before that formed part of Ethiopia
9 Then was counted as one country and consisted of Benin, Burkina Faso, Cote d'Ivoire, Guinea, Mali, Mauritania, Niger and Senegal

Table A2 - Agricultural censuses conducted in the WCA 2000 and previous WCA rounds.

Countries by Region	Year(s) of participation in the WCA round of						
	2000	1990	1980	1970	1960	1950	1930
Sierra Leone			1985	1971		1950*	
Somalia						1950* [18]	
South Africa	2000	1993		1970-71	1960	1950	1930[19]
South Sudan							
Sudan					1963		
Swaziland		1992-93	1984	1972		1950	1930
Tanzania, United Republic of	2004	1993-95		1972	1960[20]	1950[20]	
Togo	1996-97		1983	1972	1961	1950[10]*	
Tunisia	2003-04	1994-95			1961-62	1949-50	
Uganda	2002	1990-91			1963-64	1950	
Zambia	2000	1990	1982	1971	1960-61[14]	1950[21]	1929-30 [21]*
Zanzibar/Pemba[22]						1950*	
Zimbabwe					1960-61[14]	1950[23]	1929-30 [23]*
Subtotal Africa	25	23	21	25	29	18	8
AMERICA, NORTH AND CENTRAL							
Alaska[24]					1960	1950	1929
Antigua and Barbuda			1984	1973-74	1961	1950[25]*	1929-30[25]*
Bahamas		1994	1978			1950	1929-30*
Barbados		1989	1984	1971	1961	1950	1929-30
Belize			1985	1973-74		1950[26]*	
Bermuda						1950*	
Canada	1996 & 2001	1986 & 1991	1976 & 1981	1966 & 1971	1956 & 1961	1951	1931
Costa Rica			1985	1973	1963	1950	
Cuba						1952	
Dominica		1995		1974		1950[27]*	1929-30[27]*
Dominican Republic			1982	1971	1960	1950	
El Salvador				1971	1961	1950	1929
Grenada		1995	1981	1975	1961	1950[27]*	1929-30[27]*
Guadeloupe	2000-01	1989	1980	1972[28]			
Guatemala	2003		1979		1964	1950	1930
Haiti				1971		1950	
Hawaii[24]					1959	1950	1930
Honduras		1993		1974		1952	
Jamaica	1996		1978	1968-69	1961	1950	
Martinique	2000-01	1989	1980	1972[28]			
Mexico		1991	1981	1970	1960	1950	1930
Montserrat				1972		1950[25]*	1929-30[25]*
Nicaragua	2001			1971	1963		
Panama	2001	1990	1981	1971	1961	1950	
Puerto Rico	1997 & 2002	1987 & 1992	1978	1970	1959	1950	1930
Saint Kitts and Nevis	2000	1987				1950[25]*	1929-30[25]*
Saint Lucia	1996	1986	1984	1973-74		1950[27]*	1929-30[27]*
Saint Vincent and the Grenadines	2000	1986		1972-73		1950[27]*	1929-30[27]*
Trinidad and Tobago	2004		1982		1964	1951*	
United States of America	1997 & 2002	1987 & 1992	1978	1969	1959	1950	1930
Virgin Islands, United States	1998 & 2003	1988 & 1993	1978	1970	1959	1950	1930
Subtotal America, North and Central	14	16	19	23	18	18	10

10 Then part of Gold Coast and British Togoland
11 Then called Portuguese Guinea
12 Then called Basutoland
13 Then called Kingdom of Libya
14 Then part of the Federation of Rhodesia and Nyasaland
15 Then called Nyasaland and formed part of Federation of Rhodesia and Nyasaland
16 Then called South West Africa

17 Then included British Cameroon which is now part of Cameroon
18 Relates only to British Somaliland which now is part of Somalia
19 Then called Union of South Africa
20 Related to Tanganyika which now is part of United Republic of Tanzania
21 Then called Northern Rhodesia and formed part of Federation of Rhodesia and Nyasaland
22 Since 1963 forming part of United Republic of Tanzania
23 Then called Southern Rhodesia and formed part of Federation of Rhodesia and Nyasaland

Table A2 - Agricultural censuses conducted in the WCA 2000 and previous WCA rounds.

Countries by Region	Year(s) of participation in the WCA round of						
	2000	1990	1980	1970	1960	1950	1930
AMERICA, SOUTH							
Argentina	2002	1988		1969	1960	1952	1930
Bolivia, Plurinational State of					1964	1950	
Brazil	1996	1986	1980	1970	1960	1950	
Chile	1996-97		1976		1965	1955	1930
Colombia	2001-02	1988		1970-71	1960	1951	
Ecuador	1999-2000		1984	1974	1962	1954	
French Guiana	2000-01	1988-89	1980	1972			
Guyana				1968-69		1950[29]*	
Paraguay		1991	1981		1961		
Peru		1994	1983	1972	1961		1929
Suriname			1981	1969	1959		
Uruguay	2000	1990	1980	1970	1961	1951	1930
Venezuela, Bolivarian Republic of	1997-98		1985	1971	1961	1949	
Subtotal America, South	8	7	9	10	11	8	4
ASIA							
Afghanistan	2003						
Armenia[30]							
Azerbaijan[30]	2005						
Bahrain			1980	1973-74			
Bangladesh	2005	1996-97	1977				
Bhutan	2002						
Brunei Darussalam					1964	1950*	
Cambodia							
China	1997				1961[31]	1950[31] *	
Cyprus	2003-04	1994-95	1977				
Georgia[30]	2004-05						
India	1995-96 & 2000-01	1985-86 & 1990-91	1976-77 & 1980-81	1971	1960-61	1954	1929-30[32]
Indonesia	2003	1992-94	1983	1973	1963		
Iran, Islamic Republic of	2003	1988 & 1993		1974	1960		
Iraq				1971	1958	1952	
Israel			1981	1971		1950-51	
Japan	2000 & 2005	1990 &1995	1980	1970	1960	1950	1929
Jordan	1997		1983	1972		1953	
Kazakhstan[30]							
Korea, Democratic People's Republic of							
Korea, Republic of	1996 & 2001	1990	1980	1970	1961		
Kuwait				1970			
Kyrgyzstan[30]	2002-03						
Lao People's Democratic Republic	1999			1973[33]			
Lebanon	1998-99			1970	1960-62		1929-30[34]*
Malaysia	2005		1977		1960[35]	1950[35]*	1929[36]*
Maldives							
Mongolia	2000						
Myanmar	2003	1993				1953-54[37]	
Nepal	2002	1992	1981	1972	1962		

24 Since 1959 forming part of the United States of America. It is included
    separately in the 1960 round because the United States decided, for publication
    purposes, to show Alaska and Hawaii separately from the 'conterminous United
    States' as it existed before they became states
25 Then part of Leeward Islands
26 Then called British Honduras
27 Then part of Windward Islands

28 As part of French Antilles
29 Then called British Guiana
30 Since 1922 up to 1991 was part of the USSR
31 Refers to Taiwan only
32 Including areas which later formed Pakistan and, even later, Bangladesh
33 Then called Kingdom of Laos
34 Then called Lebanese Republic

Table A2 - Agricultural censuses conducted in the WCA 2000 and previous WCA rounds.

Countries by Region	Year(s) of participation in the WCA round of						
	2000	1990	1980	1970	1960	1950	1930
North Borneo[38]					1961	1950*	
Oman		1978					
Pakistan	2000	1990	1980	1972-73	1960[39]		
Philippines	2003	1992	1981	1971	1960	1948	
Qatar	2000-01						
Ryukyu Islands [40]						1951	
Sarawak[41]					1961	1950*	1929-30*
Saudi Arabia	1998-99		1982	1972			
Singapore				1973		1950*	
Sri Lanka	2002		1982	1973	1962[42]	1952[42]	1929[42]
Syrian Arab Republic	2004		1981	1970-71	1961		
Tajikistan[30]							
Thailand	2003	1993	1978		1963	1951	
Timor-Leste							
Turkey	2001	1991	1980	1970	1963-64	1950	1927
Turkmenistan[30]							
United Arab Emirates	2004						
Uzbekistan[30]							
Viet Nam	2000-01	1994			1960-61		
Yemen	2003						
Yemen, Arab Republic[43]				1983			
Yemen, People's Dem. Republic [43]				1984		1950[44]*	
Subtotal Asia	31	14	21	20	20	11	4
EUROPE							
Albania	1998	1995					1929
Andorra							
Austria	1999-2000	1990	1980	1970	1960	1950	1930
Belarus[30]							
Belgium	2000 [annual]	1990 [annual]	1979	1970	1959-60	1950	1929-30
Bosnia and Herzegovina[45]							
Bulgaria	2003			1970			
Croatia[45]	2003						
Czech Republic[46]	2000						
Czechoslovakia (former)		1990	1980	1971		1950	1930
Denmark	2000	1989	1979	1970	1959	1949	1929
Estonia[47]	2001						1929
Finland	2000	1990	1980	1969	1959	1950	1929-30
France	2000-2001	1988	1979	1970			1929-30
Germany	1999	1991 & 1995	1979	1971	1960	1949	1933
Greece	1999-2000	1991 & 1995-96	1981	1971	1961	1950	1929-30
Hungary	2000		1981	1972		1948	
Iceland	2000						1930
Ireland	2000	1991	1980	1970	1960	1949	1929
Italy	2000	1990-91	1982	1970	1961		1930
Latvia[47]	2001						1929
Lithuania[47]	2003						1930

35 Then called Federation of Malaya which now forms part of Malaysia
36 Then called Malaya which now forms part of Malaysia
37 Then called Burma
38 Now forms part of Malaysia
39 Then including present Bangladesh
40 Now forms part of Japan
41 Now forms part of Malaysia

42 Then called Ceylon
43 Now forms part of Yemen
44 Then called Aden Protectorate
45 Since 1946 up to 1991 forming part of the Socialist Federation of Yugoslavia
46 Up to 1993 forming part of Czechoslovakia
47 Since 1940 up to 1991 was part of the USSR
48 Then called Malta and Gozo

Table A2 - Agricultural censuses conducted in the WCA 2000 and previous WCA rounds.

Countries by Region	Year(s) of participation in the WCA round of						
	2000	1990	1980	1970	1960	1950	1930
Luxembourg	2000 [annual]	1990 [annual]	1980	1970	1960	1950	
Malta	2001		1979	1969	1960[48]	1950[48]	
Moldova, Republic of[49]							
Monaco							
Montenegro[50]	2003						
Netherlands	2000 [annual]	1990 [annual]	1979	1970	1960	1950	1930
Norway	1999	1989	1979	1969	1959	1949	1929
Poland	2002	1990	1980	1970	1960		
Portugal	1999-2000	1989-90	1979	1968		1952-54	
Romania	2001-2002			1970		1948	1930
Russian Federation[30]							
Saar[51]						1948	
San Marino							
Serbia[45]	2002						
Slovakia[46]	2001						
Slovenia[45]	2000	1991					
Spain	1999	1989	1982	1972	1962		1929
Sweden	1999	1990 [annual]	1981	1971	1961	1951	1932
Switzerland	2000	1990	1980	1969		1950	1929
The former Yugoslav Republic of Macedonia[45]							
Ukraine[30]							
United Kingdom	2000 [annual]	1990 [annual]	1979	1970	1960	1950	1930
Yugoslavia (Former)			1981	1969	1960	1951	1931
Subtotal Europe	33	21	22	24	17	20	23
OCEANIA							
American Samoa	1998 & 2003	1990	1980	1970	1960	1950	1930
Australia	2001	1990	1980	1971	1960	1950	1929-30
Cook Islands	2001	1988				1950*	
Fiji		1991	1978	1968	1960*	1950*	
French Polynesia		1995-96					
Guam	1998 & 2003	1988 & 1993	1978	1970	1960	1950	1930
Kiribati	2005					1950[52]*	1929-30[52]*
Marshall Islands			1980[53]	1970[53]			
Micronesia, Federated States of			1980[53]	1970[53]			
Nauru							
New Caledonia	2002	1991-92					
New Zealand	2002	1990 & 1994	1980	1972	1960	1950	1930
Niue		1989					
Northern Mariana Islands	1998&2003	1990	1980[53]	1970[53]			
Palau		1989	1980[53]	1970[53]			
Papua New Guinea					1961-62	1951	
Samoa	1999	1989				1950[54]	
Solomon Islands						1950*	
Tonga	2001		1985			1950[55]*	1929-30*
Tuvalu	2002					1950[56]*	1929-30[56]*
Vanuatu		1994	1983			1950[57]*	1929-30[57]*
Subtotal Oceania	11	13	11	9	5	6	4
TOTAL	122	94	103	111	100	81	53

49 Since 1945 up to 1991 was part of the USSR
50 Independent State since 2006
51 Now forms part of Germany
52 Refers to the Gilbert Islands that are now part of Kiribati
53 Then part of Trust Territory of the Pacific Islands

54 Then called Western Samoa
55 Then called British Solomon Islands Protectorate
56 Refers to Ellice Islands that are now part of Tuvalu
57 Then called New Hebrides

Table A3 - Time-related characteristics of reported censuses of the WCA 2000 round

Reported censuses	Census year	Time lag from the previous census (years)	Reference period	Reference day	Publication year
AFRICA (25 REPORTED CENSUSES)					
Algeria	2001	28	2000/2001	Not specified	2003
Botswana	2004	10	2003/2004	Enumeration day[1]	2007
Cape Verde	2004	16	2003/2004	Enumeration day[2]	2004
Comoros	2004	First census	2003/2004	Not specified	2005
Côte d'Ivoire	2002	28	2001	Not specified	2005
Egypt	1999-2000	9	1999/2000	25/10/2000	2001
Ethiopia	2001-2002	11	2001/2002	Not specified	2003
Gambia	2002	First census	2001/2002	Not specified	2002
Guinea	2001	6	2000/2001	Not specified	2004
Lesotho	1999-2000	10	1999/2000	Not specified	2000
Libya	1999-2000	13	1999/2000	Not specified	2001
Madagascar	2004-2005	20	2004/2005	Not specified	2005
Mali	2004-2005	19	2004/2005	Enumeration day[3]	2007
Morocco	1996	22	1995/1996	Enumeration day[4]	1998
Mozambique	2000-2001	50	2000/2001	Not specified	2001
Namibia	1996-1997	2	1996/1997	Not specified	1998
Réunion	2000	11	1999/2000	Enumeration day[5]	2002
Senegal	1998-1999	39	1998/1999	Not specified	2000
Seychelles	1998	38	1998	Not specified	n/a
South Africa	2000	7	2000	Not specified	2002
Tanzania, United Republic of	2004	9	2002/2003	1/10/2003	2006
Togo	1996-1997	14	1996	Not specified	1998
Tunisia	2003-2004	9	2003/2004	Enumeration day[6]	2005
Uganda	2002	11	2002	Enumeration day[7]	2004
Zambia	2000	10	1999/2000	Enumeration day[8]	2003
AMERICA, NORTH AND CENTRAL (14 REPORTED CENSUSES)					
Canada	2001	5	2000/2001	15/5/2000	2002
Guadeloupe	2000-2001	12	1999/2000	Enumeration day[9]	2001
Guatemala	2003	24	2002/2003	Enumeration day[10]	2004
Jamaica	1996	18	1995	16/2/1996	1996
Martinique	2000-2001	12	2000	Not specified	2001
Nicaragua	2001	30	2000/2001	Enumeration day[11]	2001
Panama	2001	11	2000/2001	22/4/2001	2002
Puerto Rico	2002	5	2002	31/12/2002	2002
Saint Kitts and Nevis	2000	13	1999/2000	Not specified	2001
Saint Lucia	1996	10	1995	Enumeration day[12]	1996
Saint Vincent and the Grenadines	2000	14	1999/2000	Not specified	2000
Trinidad and Tobago	2004	22	2003	Not specified	2004
United States of America	2002	5	2002	31/12/2002	2002
Virgin Islands, United States	2003	5	2002	31/12/2002	2005
AMERICA, SOUTH (8 REPORTED CENSUSES)					
Argentina	2002	14	2001/2001	30/6/2002	2007
Brazil	1996	10	1995/1996	31/12/1995 & 31/7/1996[13]	1996
Chile	1996-1997	19	1996/1997	Enumeration day[14]	1998
Colombia	2001-2002	14	2001	Enumeration day[15]	2002

1 Enumeration period: January to June and July to September 2004 for the traditional subsistence sector and November to December 2004 for the commercial sector
2 Enumeration period: 3 May to 31 July 2004
3 Enumeration period: June 2004 to March 2005

4 Enumeration period: 29 October 1996 to 29 March 1997
5 Enumeration period: October 2000 to February 2001
6 Enumeration period: November 2004 to March 2005
7 Enumeration period: 13 to 20 September 2002
8 Enumeration period: 16 October to 15 November 2000

Table A3 - Time-related characteristics of reported censuses of the WCA 2000 round

Reported censuses	Census year	Time lag from the previous census (years)	Reference period	Reference day	Publication year
Ecuador	1999-2000	16	1999/2000	Enumeration day[16]	2002
French Guiana	2000-2001	12	1999/2000	Enumeration day[17]	2002
Uruguay	2000	10	1999/2000	30/6/2000	2000
Venezuela, Bolivarian Republic of	1997-1998	13	1996/1997	Enumeration day[18]	1999
ASIA (29 REPORTED CENSUSES)					
Afghanistan	2003	First census	2002/2003	Not specified	2006
Azerbaijan	2005	First census	2004/2005	1/6/2005 & 1/7/2005[19]	2005
Bangladesh	2005	8	2004/2005	Enumeration day[20]	2005
Bhutan	2002	First census	2000	Not specified	2002
China	1997	First census	1996	Enumeration day[21]	1999
Cyprus	2003-2004	9	2002/2003	1/10/2003	2006
Georgia	2004-2005	First census	2003/2004	1/10/2004	2005
India	2000-01	5	2000/2001	15/10/2003[22]	2008
Indonesia	2003	9	12 months prior to enumeration[23]	Enumeration day[23]	2005
Iran, Islamic Republic of	2003	10	2002/2003	Enumeration day[24]	2003
Japan	2000	5	1999	1/12/1999 & 1/02/2000[25]	2003
Jordan	1997	14	1996	Not specified	1999
Korea, Republic of	2001	5	1999/2000	1/12/2000	2002
Kyrgyzstan	2002-2003	First census	2002/2003	Enumeration day[26]	2004
Lao People's Democratic Republic	1999	26	1998/1999	Enumeration day[27]	2000
Lebanon	1998-1999	29	1997/1998	Enumeration day[28]	2000
Malaysia	2005	28	2004/2005	Not specified	2006
Mongolia	2000	First reported census[29]	Not relevant[30]	Enumeration day[31]	2009
Myanmar	2003	10	2002/2003	Enumeration day[32]	2005
Nepal	2002	10	2001	Enumeration day[33]	2004
Pakistan	2000	10	1999/2000	Enumeration day[34]	2003
Philippines	2003	11	2002	Enumeration day[35]	2005
Qatar	2000-2001	First census	2000/2001	Not specified	2002
Saudi Arabia	1998-1999	17	1998/1999	Enumeration day[36]	1999
Sri Lanka	2002	20	2002	Not specified	2002
Thailand	2003	10	2002/2003	1/5/2003	2003
Turkey	2001	10	2000/2001	Not specified	2001
Viet Nam	2000-2001	7	2000/2001	Enumeration day[37]	2003
Yemen	2003	19	2002	Enumeration day[38]	n/a

9 Enumeration period: October 2000 to January 2001
10 Enumeration period: 4 to 31 May 2003
11 Enumeration period: March 2001
12 Enumeration period: 15 January to 15 March 1996
13 The first reference day is for data on ownership, area and employment while the second one for data on livestock, permanent workers and forestry
14 Enumeration period: for statistical regions I,XI and XII 1January to 30 June 1996; for statistical regions II to X March 31 to May 30 1997
15 Enumeration period: June 2001 to April 2002
16 Enumeration period: 1 October 1999 to 30 September 2000
17 Enumeration period: June 2000 to January 2001
18 Enumeration period: July 1997 to January 1998

19 The first reference day is for agricultural enterprises while the second one for peasant farms
20 Enumeration period: 17 to 31 May 2005
21 Enumeration period: 1 to 31 January 1997
22 The reference day is for a livestock census conducted in 2003;
23 Enumeration period: 1 September 2002 to 31 August 2003
24 Enumeration period: October 2003
25 The first reference day is for Okinawa prefecture while the second one for the rest of Japan
26 Enumeration period: 1 to 15 June 2002 in southern part and 1 to 15 August 2002 in northern part at the first stage; 1 to 15 November 2003 at the second stage
27 Enumeration period: 22 February to 19 March 1999

Table A3 - Time-related characteristics of reported censuses of the WCA 2000 round

Reported censuses	Census year	Time lag from the previous census (years)	Reference period	Reference day	Publication year
EUROPE (29 REPORTED CENSUSES)					
Albania	1998	3	1997/1998	Enumeration day[39]	2000
Austria	1999-2000	10	1999	1/6/1999& 1/12/1999[40]	2003
Belgium	2000	1	1999/2000	15/5/2000 & 1/12/2000[41]	2001
Croatia	2003	22	2002/2003	1/6/2003	2003
Czech Republic	2000	10	1999/2000	Enumeration day[42]	2000
Denmark	2000	11	1999/2000	2/6/2000	2003
Estonia	2001	72	2001	Not specified	2001
Finland	2000	10	1999/2000	1/5/2000 & 26/5/2000[43]	2003
France	2000-2001	13	1999/2000	Enumeration day[44]	2003
Germany	1999	4	1999/2000	3/5/1999	2002
Greece	1999-2000	4	1998/1999	30/9/1999 & 1/11/1999[45]	2003
Hungary	2000	19	1999/2000	31/3/2000	2002
Ireland	2000	9	1999/2000	1/6/2000	2000
Italy	2000	9	1999/2000	22/10/2000	2000
Latvia	2001	72	2001	1/6/2001	2003
Lithuania	2003	73	2003	1/6/2003	2003
Luxembourg	2000	1	1998/1999	15/5/1999	2000
Malta	2001	22	2000/2001	30/9/2000	2003
Netherlands	2000	1	1999	1/4/1999	2000
Norway	1999	10	1999	31/7/1999	2000
Poland	2002	12	2001/2002	Not specified	2002
Portugal	1999-2000	10	1998/1999	Not specified	2001
Romania	2001-2002	32	2001/2002	1/12/2002	2004
Serbia	2002	21	2001/2002	31/3/2002	2002
Slovakia	2001	11	2000/2001	31/10/2001	2001
Slovenia	2000	9	1999/2000	1/6/2000	2000
Spain	1999	10	1998/1999	30/11/1999	1999
Sweden	1999	4	1998/1999	10/6/1999	2000
United Kingdom	2000	1	1999/2000	5/6/2000	2000
OCEANIA (9 REPORTED CENSUSES)					
American Samoa	2003	5	2002	Enumeration day[46]	2005
Australia	2001	11	2000/2001	30/6/2001	2002
Cook Islands	2001	13	12 months prior to enumeration[46]	Enumeration day[47]	2001

28 Enumeration period: 1 November 1998 to 30 July 1999
29 Mongolia has been conducting annual livestock censuses since 1924 but the census of 2000 reported in here is the first one that entered into FAO reviews. Thus it is marked as 'First reported census'.
30 This was a livestock census
31 Enumeration period: 7 to 17 December 2000
32 Enumeration period: 1 October to 15 November 2003
33 Enumeration period: January to June 2002 in 46 districts and April to June 2002 in the remaining 29 districts
34 Enumeration period: January to March 2000 for 'hot area'; April to June 2000 for 'cold area'; July to August 2000 for 'very cold area' northern areas)

35 Enumeration period: 3 March to 5 April 2003
36 Enumeration period not specified
37 Enumeration period: 1 to 30 October 2001
38 Enumeration period: April to May 2003
39 Data collection started on 1 June 1998
40 The first reference day is for farm structure survey while the second one for livestock numbers
41 The first reference day is for land use, livestock and agricultural machinery while the second one for farm labour force
42 Enumeration period: October to December 2000
43 The first reference day is for livestock while the second one for land use
44 Enumeration period: 1 October 2000 to 31 January 2001

Table A3 - Time-related characteristics of reported censuses of the WCA 2000 round

Reported censuses	Census year	Time lag from the previous census (years)	Reference period	Reference day	Publication year
Guam	2003	5	2002	Enumeration day[48]	2004
New Caledonia	2002	10	2002	Not specified	2003
New Zealand	2002	8	2001/2002	30/6/2002	2002
Northern Mariana Islands	2003	5	2002	Enumeration day[49]	2003
Samoa	1999	10	1999	Enumeration day[50]	2000
Tonga	2001	16	2000/2001	Enumeration day[51]	2001

45 The first reference day is for machinery while the second one is for livestock
46 Enumeration period: January to February 2003
47 Enumeration period: 13 November to 13 December 2001

48 Enumeration period: spring season of 2003
49 Enumeration period: January 2003
50 Enumeration period: 18 October to 30 November 1999
51 Enumeration period: mid-December 2001

Table A4 - Enumeration methods and sample designs used in the WCA 2000 round

| Reported Censuses | ENUMERATION METHOD | | | SAMPLE DESIGN | | |
| | Complete Enumeration | Sample Enumeration | One stage | Multiple stages (in brackets – selection method) | | |
				PSU	SSU	TSU
AFRICA (25 REPORTED CENSUSES)						
Algeria	✓					
Botswana[1]	✓	✓		EA (pps)	AH (ep)	
Cape Verde	✓					
Comoros		✓		V (pps)	AH (ep)	
Cote d'Ivoire[2]	✓	✓		EA (pps)	AH (pps)	
Egypt	✓					
Ethiopia		✓		EA (pps)	HH (pps)	
Gambia		✓		EA (pps)	HH (ep)	
Guinea		✓		EA (pps)	AH (ep)	
Lesotho		✓		EA (pps)	AH (ep)	
Libya	✓					
Madagascar		✓		V (ep)	AH (ep)	
Mali[3]	✓	✓		EA (pps)	AH (ep)	
Morocco[4]	✓	✓	NA	NA	NA	NA
Mozambique[5]	✓	✓		V (ep)	AH (ep)	
Namibia		✓		EA (na)	AH (na)	
Réunion	✓					
Senegal		✓		EA (pps)	AH (ep)	
Seychelles	✓					
South Africa		✓		EA (ep)	AH (ep)	
Tanzania, United Republic of[6]	✓	✓		EA (pps)	AH (ep)	
Togo		✓		EA (pps)	AH (ep)	
Tunisia[7]	✓	✓		EA (na)	AH (p1)	
Uganda	✓					
Zambia	✓					
AMERICA, NORTH AND CENTRAL (14 REPORTED CENSUSES)						
Canada	✓					
Guadeloupe	✓					
Guatemala	✓					
Jamaica[8]	✓	✓	✓			
Martinique	✓					
Nicaragua	✓					
Panama	✓					
Puerto Rico	✓					
Saint Kitts and Nevis	✓					
Saint Lucia	✓					
Saint Vincent and the Grenadines	✓					
Trinidad and Tobago	✓					
United States of America[9]	✓	✓	✓			
Virgin Islands, United States	✓					

1 Complete enumeration for the Commercial Sector; sample enumeration for the Traditional Subsistence Sector
2 Complete enumeration of the Modern Sector and large holdings of Traditional Sector; sample enumeration for the remaining holdings of the Traditional Sector
3 Complete enumeration of modern holdings; sample enumeration for the Traditional Sector
4 A short questionnaire applied to all holdings and a long questionnaire to a sample
5 Complete enumeration of large holdings and sample enumeration of small and medium holdings
6 Complete enumeration for large scale farms; sample enumeration for others
7 Complete enumeration of large scale holdings; two-stage sampling (one-stage cluster sampling) applied to medium and small-scale holdings with sampling EAs at the first stage and enumerating all holdings in EAs at the second stage
8 10 % sample survey of farms with less than 50 acres; complete enumeration of all other farms
9 'Non-Sample Form' collecting basic information was applied to all holdings while 'Sample Form' collecting additional information was applied to a sample of holdings selected using one-stage sampling.

Reported Censuses	ENUMERATION METHOD			SAMPLE DESIGN		
	Complete Enumeration	Sample Enumeration	One stage	Multiple stages (in brackets – selection method)		
				PSU	SSU	TSU
AMERICA, SOUTH (8 REPORTED CENSUSES)						
Argentina	✓					
Brazil	✓					
Chile	✓					
Colombia[10]	✓	✓		PFU	S	
Ecuador[11]	✓	✓		PFU	S	
French Guiana	✓					
Uruguay	✓					
Venezuela, Bolivarian Republic of	✓					
ASIA (29 REPORTED CENSUSES)						
Afghanistan[12]	✓	✓		D/V (na)	AH (na)	
Azerbaijan[13]	✓	✓	NA	NA	NA	NA
Bangladesh		✓		V (pps)	AH (ep)	
Bhutan	✓					
China	✓					
Cyprus	✓					
Georgia	✓					
India[14]	✓	✓		V (ep)	AH (p1)	
Indonesia		✓		EA (pps)	AH (ep)	
Iran, Islamic Republic of	✓					
Japan	✓					
Jordan	✓					
Korea, Republic of	✓					
Kyrgyzstan[15]	✓	✓		O (pps)	HH (ep)	
Lao, People's Democratic Republic[16]	✓	✓		V (pps)	HH (ep)	
Lebanon	✓					
Malaysia	✓					
Mongolia	✓					
Myanmar	✓					
Nepal		✓		EA (pps)	AH (ep)	
Pakistan[17]		✓		D (pps)	V (pps)	HH clusters ep
Philippines[18]		✓		V (ep)	HH (p1)	
Qatar	✓					
Saudi Arabia	✓					
Sri Lanka	✓					
Thailand	✓					
Turkey[19]	✓	✓		V (pps)	AH (ep)	
Viet Nam[20]	✓	✓	NA	NA	NA	NA
Yemen[21]	✓	✓		EA (na)	AH (na)	

▶

Alaska and Rhode Island were covered only by the sampling method; complete enumeration was applied to holdings with special characteristics (e.g. with large total value of agricultural products or large area) and counties with less than 100 holdings

10 Complete enumeration for special crops and two-stage area sample design for other holdings

11 Complete enumeration for large and specialized holdings and two-stage area sample design for other holdings

12 Complete enumeration of livestock holdings with basic questionnaire; sample enumeration for obtaining additional information

13 Complete enumeration of all institutional farms; sample enumeration of peasant farms and homesteads; no information available as to sampling design

14 Complete enumeration of all holdings for collecting livestock data and basic data for crops; two-stage sample enumeration one-stage cluster sampling for more detailed information on crops with 20% of villages selected at the first stage and all holdings enumerated in the selected villages

15 Sample enumeration was applied to household plots. The primary sampling unit was a street in urban area and a household book in rural area in rural area households are registered in household books; there may be more than one household book in a village)

16 Complete enumeration for collecting some basic data; sample enumeration for detailed information

17 Three stage weighted and stratified cluster sampling design for rural settled areas of North West Frontier Provinces, Punjab and Sindh

Table A4 - Enumeration methods and sample designs used in the WCA 2000 round

Reported Censuses	ENUMERATION METHOD			SAMPLE DESIGN		
	Complete Enumeration	Sample Enumeration	One stage	Multiple stages (in brackets – selection method)		
				PSU	SSU	TSU
EUROPE (29 REPORTED CENSUSES)						
Albania	✓					
Austria	✓					
Belgium	✓					
Croatia	✓					
Czech Republic	✓					
Denmark	✓					
Estonia	✓					
Finland	✓					
France	✓					
Germany[22]	✓					
Greece	✓					
Hungary	✓					
Ireland	✓					
Italy	✓					
Latvia	✓					
Lithuania	✓					
Luxembourg	✓					
Malta	✓					
Netherlands	✓					
Norway	✓					
Poland	✓					
Portugal	✓					
Romania	✓					
Serbia	✓					
Slovakia[23]	✓	✓	✓			
Slovenia	✓					
Spain	✓					
Sweden	✓					
United Kingdom[24]	✓	✓	✓			
OCEANIA (9 REPORTED CENSUSES)						
American Samoa	✓					
Australia	✓					
Cook Islands	✓					
Guam	✓					
New Caledonia	✓					
New Zealand		✓	✓			
Northern Mariana Islands	✓					
Samoa[25]	✓	✓	✓			
Tonga	✓					

provinces at the third stage clusters of households were selected and all of those households – in fact, fourth order sampling units – were enumerated); single stage weighted sampling for rural settled area of Baluchistan, Azad Jammu and Kashmir; single stage systematic sampling in the rest of the country

18 Two-stage sampling (one-stage cluster sampling) with 28.4% of barangays (PSUs) of the country selected at the first sta ge and all holdings (SSUs) within selected barangays listed and interviewed.

19 Complete enumeration for villages and settlements with 5 000 inhabitants and more; sample enumeration for holdings in villages and settlements with less than 5 000 inhabitants

20 No information is available as to complete and sample enumeration details and sampling design

21 Complete enumeration refers to prelisting activities collecting basic agricultural information for all holdings; the main census questionnaire was applied to a sample of holdings

22 Special sample surveys focused on activities like wine growing, market gardening and inland water fishing was conducted together with the census

23 Complete enumeration of holdings above a given threshold; sample enumeration of holdings below the threshold

24 Sampling of holdings with a standard Gross Margin of less than 9 600

25 The complete enumeration of all households concerned the first part of the questionnaire while the second part was designed to cover 25% of the agricultural holdings

Table A5 - Enumeration Techniques used during the WCA 2000 round

Reported censuses by region	ENUMERATION TECHNIQUES			Objective measurement used (A=Area; Y=Yield)
	Interview (a)	Mail (b)	Combination of (a) and (b)	
AFRICA (25 REPORTED CENSUSES)				
Algeria	✓			
Botswana			✓	
Cape Verde	✓			
Comoros	✓			A&Y
Cote d'Ivoire	✓			
Egypt	✓			
Ethiopia	✓			A&Y
Gambia	✓			A&Y
Guinea	✓			
Lesotho	✓			A&Y
Libyan Arab Jamahiriya	✓			
Madagascar	✓			A&Y
Mali	✓			
Morocco	✓			
Mozambique	✓			A
Namibia	✓			
Réunion	✓			
Senegal	✓			A&Y
Seychelles	✓			
South Africa	✓			
Tanzania, United Republic of	✓			
Togo	✓			A&Y
Tunisia	✓			
Uganda	✓			
Zambia	✓			
Subtotal Africa	**24**	**0**	**1**	**8**
AMERICA, NORTH AND CENTRAL (14 REPORTED CENSUSES)				
Canada			✓[1]	
Guadeloupe	✓			
Guatemala	✓			
Jamaica	✓			
Martinique	✓			
Nicaragua	✓			
Panama	✓			
Puerto Rico			✓[2]	
Saint Kitts and Nevis	✓			
Saint Lucia	✓			
Saint Vincent and the Grenadines	✓			
Trinidad and Tobago	✓			
United States of America			✓[3]	
Virgin Islands, United States	✓			

1 For questionnaires not received by mail and for missing data, a follow-up was conducted by telephone or face-to face interview
2 The farms not responding by mail were visited by enumerators
3 Approximately 30 000 respondents were selected by field offices for personal enumeration rather than mail out/mail back enumeration
4 Small holdings visited by enumerators, large holdings enumerated by mail questionnaire

Table A5 - Enumeration Techniques used during the WCA 2000 round

Reported censuses by region	ENUMERATION TECHNIQUES			Objective measurement used (A=Area; Y=Yield)
	Interview (a)	Mail (b)	Combination of (a) and (b)	
Subtotal America, North and Central	11	0	3	0
AMERICA, SOUTH (8 REPORTED CENSUSES)				
Argentina	✓			
Brazil	✓			
Chile	✓			
Colombia	✓			A&Y
Ecuador	✓			
French Guiana	✓			
Uruguay	✓			
Venezuela, Bolivarian Republic of	✓			
Subtotal America, South	8	0	0	1
ASIA (29 REPORTED CENSUSES)				
Afghanistan	✓			
Azerbaijan	✓			
Bangladesh	✓			
Bhutan	✓			
China	✓			
Cyprus	✓			
Georgia	✓			
India	✓			
Indonesia	✓			
Iran, Islamic Republic of	✓			
Japan	✓			
Jordan	✓			
Korea, Republic of	✓			
Kyrgyzstan	✓			A
Lao People's Democratic Republic	✓			A
Lebanon	✓			
Malaysia	✓			
Mongolia	✓			
Myanmar	✓			
Nepal	✓			
Pakistan	✓			
Philippines	✓			
Qatar	✓			A
Saudi Arabia	✓			
Sri Lanka			✓[4]	
Thailand	✓			
Turkey	✓			
Viet Nam	✓			
Yemen	✓			
Subtotal Asia	28	0	1	3

5 Households were enumerated by direct interview; business entities were enumerated by mail
6 A few thousand of major holdings were surveyed by mail
7 Questionnaires sent by mail and telephonic interviews carried out
8 Cooperatives and companies investigated by mail

Table A5 - Enumeration Techniques used during the WCA 2000 round

Reported censuses by region	ENUMERATION TECHNIQUES			Objective measurement used (A=Area; Y=Yield)
	Interview (a)	Mail (b)	Combination of (a) and (b)	
EUROPE (29 REPORTED CENSUSES)				
Albania	✓			
Austria	✓			
Belgium	✓			
Croatia			✓5	
Czech Republic			✓6	A
Denmark		✓		
Estonia	✓			
Finland			✓7	
France	✓			
Germany	✓			
Greece	✓			
Hungary			✓8	
Ireland		✓		
Italy	✓			
Latvia	✓			
Lithuania	✓			
Luxembourg	✓			
Malta			✓9	
Netherlands			✓2	
Norway		✓		
Poland	✓			
Portugal	✓			
Romania	✓			
Slovakia	✓			
Serbia	✓			
Slovenia			✓10	
Spain	✓			
Sweden		✓		
United Kingdom		✓		
Subtotal Europe	**17**	**5**	**7**	**1**
OCEANIA (9 REPORTED CENSUSES)				
American Samoa	✓			
Australia			✓2	
Cook Island	✓			
Guam	✓			
New Caledonia	✓			
New Zealand		✓		
Northern Mariana Islands	✓			
Samoa	✓			
Tonga	✓			
Subtotal Oceania	**7**	**1**	**1**	**0**
TOTAL	**96**	**6**	**12**	**13**

9 Holdings having less than 3 ha of dry land and no means of irrigation were surveyed by mail; all others were enumerated by personal interviews

10 Data on agricultural enterprises collected by mail

Table A6 - Census Frames used in the WCA 2000 round

Reported Censuses by Region	Frame type for complete enumeration (the whole census or a component)	Frame type for sample enumeration (the whole census or a component)	
		One-stage	Multiple-stage
AFRICA (25 REPORTED CENSUSES)			
Algeria	Prelisting activities		
Botswana	Administrative sources		List of AH from a PC
Cape Verde	Screening EAs		
Comoros			Screening selected villages
Cote d'Ivoire	Adm. & non-adm. sources		List of AH from a PC
Egypt	Prelisting activities		
Ethiopia			Screening selected EAs
Gambia			Screening selected EAs
Guinea			Screening selected EAs
Lesotho			Screening selected EAs
Libya	List of AH from a PC		
Madagascar			Screening selected villages
Mali	Adm. & non-adm. sources		Screening selected EAs
Morocco	NA		NA
Mozambique	Administrative sources		Screening selected villages
Namibia			Non-administrative sources
Réunion	Prelisting activities		
Senegal			Prelisting activities
Seychelles	Population census module		
South Africa			Screening selected EAs
Tanzania, United Republic of	Adm. & non-adm. sources		Screening selected EAs
Togo			Prelisting activities
Tunisia	Adm. & non-adm. sources		Screening selected EAs
Uganda	Population census module		
Zambia	Population census module		
AMERICA, NORTH AND CENTRAL (14 REPORTED CENSUSES)			
Canada	Maintained farm register[1]		
Guadeloupe	Adm. & non-adm. sources		
Guatemala	Screening EAs		
Jamaica	Prelisting activities	Prelisting activities	
Martinique	Prelisting activities		
Nicaragua	List of HH from a PC		
Panama	Screening EAs		
Puerto Rico	Adm. & non-adm. sources		Area frame
Saint Kitts and Nevis	Screening EAs		
Saint Lucia	Adm. & non-adm. sources		
Saint Vincent	Screening EAs		
Trinidad and Tobago	List of AH from a PC		
United States of America	Adm. & non-adm. sources	Adm. & non-adm. sources	
Virgin Islands, United States	Screening EAs		
AMERICA, SOUTH (8 REPORTED CENSUSES)			
Argentina	Non-administrative sources		
Brazil	Screening EAs		
Chile	Screening EAs		
Colombia	Administrative sources		Area frame
Ecuador	Adm. & non-adm. sources		Area frame
French Guiana	Prelisting activities		
Uruguay	Screening EAs		
Venezuela, Bolivarian Republic of	Screening EAs		

1 The Agricultural Census was conducted jointly with the Population Census. The enumerators, having delivered the Population Census Questionnaire, delivered the Agricultural Census Questionnaire as well if anyone in the household operated a farm, as indicated by the Farm Register maintained by Statistics Canada.

Table A6 - Census Frames used in the WCA 2000 round

Reported Censuses by Region	Frame type for complete enumeration (the whole census or a component)	Frame type for sample enumeration (the whole census or a component)	
		One-stage	Multiple-stage
ASIA (29 REPORTED CENSUSES)			
Afghanistan	Screening EAs		Prelisting activities[2]
Azerbaijan	Administrative sources		NA
Bangladesh			Administrative sources
Bhutan	Screening districts		
China	List of HH from a PC		
Cyprus	Adm. & non-adm. sources		
Georgia	Screening EAs		
India	Administrative sources		Administrative sources
Indonesia			Prelisting activities
Iran, Islamic Republic of	Screening EAs		
Japan	Prelisting activities		
Jordan	List of HH from a PC		
Korea, Republic of	Screening EAs		
Kyrgyzstan	Administrative sources		
Lao People's Democratic Republic	List of HH from a PC		List of HH from a PC
Lebanon	Prelisting activities		
Malaysia	Screening EAs		
Mongolia	Screening villages		
Myanmar	Prelisting activities		
Nepal			List of AH from a PC[3]
Pakistan			Administrative sources
Philippines			Screening selected villages
Qatar	Prelisting activities		
Saudi Arabia	Screening villages		
Sri Lanka	List of HH from a PC		
Thailand	Screening EAs		
Turkey	Administrative sources		Screening selected villages
Viet Nam	Administrative sources		NA
Yemen	Screening EAs		Prelisting activities[4]
EUROPE (29 REPORTED CENSUSES)			
Albania	Screening EAs		
Austria	Maintained farm register		
Belgium	Maintained farm register		
Croatia	List of AH from a PC		
Czech Republic	Administrative sources		
Denmark	Maintained farm register		
Estonia	Prelisting activities		
Finland	Administrative sources		
France	Adm. & non-adm. sources		
Germany	Adm. & non-adm. sources		
Greece	Screening EAs		
Hungary	Screening EAs		
Ireland	Adm. & non-adm. sources		
Italy	Adm. & non-adm. sources		
Latvia	Adm. & non-adm. sources		
Lithuania	Screening EAs		

2 The complete enumeration component of the census was conducted during the first phase where enumeration areas were screened for agricultural holdings and basic information was collected. This can be considered as pre-listing activities for the creation of the frame for the sample component of the census.

3 Some additional questions on areas and numbers of livestock and poultry were asked during the listing operation of the 2001 Population Census which also enabled the preparation of the frame for the National Sample Census of Agriculture

Table A6 - Census Frames used in the WCA 2000 round

Reported Censuses by Region	Frame type for complete enumeration (the whole census or a component)	Frame type for sample enumeration (the whole census or a component)	
		One-stage	Multiple-stage
Luxembourg	Screening communes		
Malta	Adm. & non-adm. sources		
Netherlands	Maintained farm register		
Norway	Maintained farm register		
Poland	Adm. & non-adm. sources		
Portugal	Prelisting activities		
Romania	Adm. & non-adm. sources		
Serbia	Population census module		
Slovakia	Adm. & non-adm. sources	Adm. & non-adm. sources	
Slovenia	Adm. & non-adm. sources		
Spain	Screening EAs		
Sweden	Maintained farm register		
United Kingdom	Maintained farm register	Maintained farm register	
OCEANIA (9 REPORTED CENSUSES)			
American Samoa	Adm. & non-adm. sources		
Australia	Maintained farm register		
Cook Islands	Screening EAs		
Guam	Administrative sources		
New Caledonia	Screening EAs		
New Zealand		Maintained farm register	
Northern Mariana Islands	Administrative sources		
Samoa	Prelisting activities	Prelisting activities	
Tonga	Screening EAs		

4 The complete enumeration component of the census was conducted during the listing stage where about 1.3 million holdings were identified on the basis of basic information on agricultural land, type of irrigation used and livestock inventories. This can be considered as pre-listing activities for creation of the frame for the sample component of the census.

Table A7 - Geographic and holdings coverage of reported censuses of the 2000 WCA round

Reported censuses by Region	Geographical coverage	Holdings coverage
AFRICA[2] (5 REPORTED CENSUSES)		
Algeria	Entire country	All holdings
Botswana	Urban areas not covered	All holdings
Cape Verde	Entire country	All holdings
Comoros	Entire country	All holdings
Cote d'Ivoire	Entire country	All holdings
Egypt	Entire country	Holdings below a threshold excluded
Ethiopia	Some areas not covered[1]	All holdings
Gambia	Entire country	All holdings
Guinea	Entire country	All holdings
Lesotho	Entire country	All holdings
Libyan Arab Jamahiriya	Entire country	All holdings
Madagascar	Entire country	All holdings
Mali	Entire country	All holdings
Morocco	Entire country	All holdings
Mozambique	Some areas not covered[2]	All holdings
Namibia	Some areas not covered[3]	All holdings
Réunion	Entire country	Holdings below a threshold excluded
Senegal	Some areas not covered[4]	Some holdings excluded[5]
Seychelles	Entire country	All holdings
South Africa	Entire country	Some holdings excluded[5]
Tanzania, United Republic Of	Urban areas not covered	Holdings below a threshold excluded
Togo	Entire country	All holdings
Tunisia	Entire country	Holdings below a threshold excluded
Uganda	Entire country	Some holdings excluded[7]
Zambia	Entire country	All holdings
AMERICA, NORTH AND CENTRAL (14 REPORTED CENSUSES)		
Canada	Entire country	Some holdings excluded[8]
Guadeloupe	Entire country	Holdings below a threshold excluded
Guatemala	Some areas not covered[9]	All holdings
Jamaica	Entire country	Holdings below a threshold excluded
Martinique	Entire country	Holdings below a threshold excluded
Nicaragua	Entire country	Some holdings excluded[10]
Panama	Entire country	All holdings
Puerto Rico	Entire country	All holdings
Saint Kitts and Nevis	Entire country	Holdings below a threshold excluded
Saint Lucia	Entire country	Holdings below a threshold excluded
Saint Vincent	Entire country	Holdings below a threshold excluded
Trinidad and Tobago	Entire country	Holdings below a threshold excluded
United States	Entire country	Holdings below a threshold excluded
Virgin Islands, United States	Entire country	Holdings below a threshold excluded
AMERICA, SOUTH (8 REPORTED CENSUSES		
Argentina	Some areas not covered[11]	Holdings below a threshold excluded
Brazil	Entire country	Some holdings excluded[12]
Chile	Entire country	Holdings below a threshold excluded

1 Pastoralist areas of the Afar and Somali Regional States not covered
2 Cities of Maputo, Matola, Beira and Nampula (investigated separately) excluded; some districts in the province of Zambezia were also excluded due to adverse natural events
3 Only 6 Regions (Ohamgwena, Omusati, Oshana, Oshikoto, Kawango and Caprivi) out of 13, making up the Northern Communal Areas of Namibia, were covered.
4 Urban areas were not covered; the departments of Gignona,

Oussouye and Ziguinchor, and the district of Diattacounda were also not covered for insecurity reasons
5 Holdings practicing the rainfall (winter) agriculture were covered
6 Small-scale market gardens in peri-urban areas not covered.
7 No investigation of private large scale and institutional farm sector
8 Only holdings producing some agricultural products for sale covered
9 For security reasons
10 Kitchen-gardens in urban areas excluded

Table A7 - Geographic and holdings coverage of reported censuses of the 2000 WCA round

Reported censuses by Region	Geographical coverage	Holdings coverage
Colombia	Some areas not covered[13]	All holdings
Ecuador	Entire country	Holdings below a threshold excluded
French Guiana	Entire country	Holdings below a threshold excluded
Uruguay	Entire country	Holdings below a threshold excluded
Venezuela	Entire country	Holdings below a threshold excluded[14]
ASIA (29 REPORTED CENSUSES)		
Afghanistan	Some areas not covered[15]	Only livestock holdings covered
Azerbaijan	Occupied lands excluded	All holdings
Bangladesh	Entire country	Holdings below a threshold excluded
Bhutan	Entire country	All holdings
China	Entire country	All holdings
Cyprus	Some areas not covered[16]	Holdings below a threshold excluded
Georgia	Some areas not covered[18]	Some holdings excluded[18]
India	Some areas not covered[19]	All holdings
Indonesia	Entire country	All holdings
Iran, Islamic Republic of	Entire country	Holdings below a threshold excluded[20]
Japan	Entire country	All holdings
Jordan	Entire country	Holdings below a threshold excluded
Korea, Rep. of	Entire country	Holdings below a threshold excluded
Kyrgyzstan	Entire country	All holdings
Lao People's Democratic Republic	Entire country	Holdings below a threshold excluded
Lebanon	Entire country	Holdings below a threshold excluded
Malaysia	Entire country	Some holdings excluded[21]
Mongolia	Entire country	Only livestock holdings covered
Myanmar	Some areas not covered[22]	Holdings below a threshold excluded
Nepal	Entire country	Holdings below a threshold excluded[23]
Pakistan	Entire country	Holdings below a threshold excluded
Philippines	Entire country	Holdings below a threshold excluded
Qatar	Entire country	Holdings below a threshold excluded
Saudi Arabia	Some areas not covered[24]	All holdings
Sri Lanka	Some areas not covered[25]	Holdings below a threshold excluded
Thailand	Entire country	All holdings
Turkey	Entire country	All holdings
Viet Nam	Entire country	All holdings
Yemen	Entire country	All holdings
EUROPE (29 REPORTED CENSUSES)		
Albania	Entire country	All holdings
Austria	Entire country	Holdings below a threshold excluded
Belgium	Entire country	Some holdings excluded[26]
Croatia	Entire country	All holdings
Czech Republic	Entire country	Holdings below a threshold excluded
Denmark	Entire country	Holdings below a threshold excluded
Estonia	Entire country	Holdings below a threshold excluded
Finland	Entire country	Holdings below a threshold excluded
France	Entire country	Holdings below a threshold excluded

11 Urban areas and desert zones with no agricultural use not covered
12 Holdings consisting of only family gardens and forest holdings excluded
13 Urban, peri-urban areas and non-agricultural regions not covered
14 Minimum size for inclusion not specified
15 Barmal District and parts of Ghor Province not covered for security reasons

16 Only the area controlled by Government of Cyprus covered
17 Uncontrolled territories in Abkhazia and Tskhinvali Region not covered
18 Holdings whose holders lived or had headquarters (in case of juridical persons) in 5 large cities (Tbilisi, Kutaisi, Batumi, Rustavi and Poti) were not covered
19 States of Bihar, Jharkhand and Megalaya not covered. No data for Livestock Census in some districts. In some states urban areas were excluded

▶ Table A7 - Geographic and holdings coverage of reported censuses of the 2000 WCA round

Reported censuses by Region	Geographical coverage	Holdings coverage
Germany	Entire country	Holdings below a threshold excluded
Greece	Entire country	Holdings below a threshold excluded
Hungary	Entire country	Holdings below a threshold excluded
Ireland	Entire country	Holdings below a threshold excluded
Italy	Entire country	All holdings
Latvia	Entire country	Holdings below a threshold excluded
Lithuania	Entire country	Holdings below a threshold excluded
Luxembourg	Entire country	Holdings below a threshold excluded[27]
Malta	Entire country	All holdings
Netherlands	Entire country	Holdings below a threshold excluded
Norway	Entire country	Holdings below a threshold excluded
Poland	Entire country	Holdings below a threshold excluded
Portugal	Entire country	Holdings below a threshold excluded
Romania	Entire country	All holdings
Serbia	Entire country	Holdings below a threshold excluded[28]
Slovakia	Entire country	All holdings
Slovenia	Entire country	Holdings below a threshold excluded
Spain	Entire country	Holdings below a threshold excluded
Sweden	Entire country	Holdings below a threshold excluded
United Kingdom	Entire country	All holdings
OCEANIA (9 REPORTED CENSUSES)		
American Samoa	Entire country	All holdings
Australia	Entire country	Holdings below a threshold excluded
Cook Islands	Entire country	Holdings below a threshold excluded[29]
Guam	Entire country	Holdings below a threshold excluded
New Caledonia	Entire country	Holdings below a threshold excluded
New Zealand	Entire country	All holdings
Northern Mariana Islands	Entire country	Holdings below a threshold excluded
Samoa	Entire country	All holdings
Tonga	Entire country	All holdings

20 Along with holdings below a threshold, *modern poultry farms were also not covered*
21 Only crop holdings covered
22 Highly urbanized areas and those facing security problems not covered
23 Along with holdings below a threshold, *holdings operated by government organizations, corporations and other juridical persons were also not covered*
24 Agriculturally productive regions covered
25 Municipalities of Colombo, Dehiwala, Mount Lavinia and Sri Jaywardanepura, predominantly used for residential and commercial purposes, not covered

26 Criteria for inclusion: at least 1 ha of land or production for sale
27 Criteria for inclusion: at least 1 ha of cultivated area or 0.1 ha of vineyards; or commercial units producing vegetables, fruit or flowers; or owners of nurseries or osieries; or commercial livestock or poultry breeding establishments
28 Along with holdings below a threshold, holdings operated by agricultural enterprises and cooperatives (legal units) were also not covered
29 Along with holdings below a threshold, *holdings operated by government, communities and institutions were also not covered*

Table A8 - Minimum thresholds for inclusion of holdings used in the WCA 2000 round

Censuses by Region	Land	Trees	Cattle/buffaloes/camels	Sheep	Goats	Pigs	Poultry	Other livestock	Value/sales	Other
AFRICA (4 CENSUSES WITH REPORTED MINIMUM THRESHOLDS)										
Egypt	87.5 sq. m.(1) of agricultural land		1 head of cattle; or 1 buffalo; or 1 camel	5 sheep/goats in total			100	10 beehives		1 fishery cage; or 1 agricultural machine owned/shared
Reunion	1 ha of agricultural land; or 0.2 ha under specialized crops; or 0.1 ha under bananas; or 0.1 ha under sugarcane		10 cows					10 beehives		
Tanzania, United Republic Of	25 sq. m. of arable land		1 head of cattle	5 sheep/goats/pigs in total			50 chickens/ ducks/poultry in total			
Tunisia	0.01 ha under irrigated cultures; or 1 ha under non-irrigated cultures		1 milk cow; or 10 female camels	6 female sheep/ goats in total			500 chickens and 50 laying hens	10 mares; or 50 adult female-rabbits, 5 cages of adult male-rabbits and 5 cages of young rabbits		
AMERICA, NORTH AND CENTRAL (9 CENSUSES WITH REPORTED MINIMUM THRESHOLDS)										
Guadeloupe	1 ha of agricultural land; or 0.2 ha under specialized crops; or 0.1 ha under bananas; or 0.1 ha under sugarcane		10 cows					10 beehives		
Jamaica	0.41 ha(2) under crops; or 409 sq. m.(3) of greenhouses	12 Economic trees like citrus, mangoes, breadfruit etc.	1 head of cattle	2 sheep/goats/pigs in total			12	6 beehives		1 fish pond
Martinique	1 ha of agricultural land; or 0.2 ha under specialized crops; or 0.1 ha under bananas; or 0.1 ha under sugarcane		10 cows					10 beehives		
Saint Kitts and Nevis	0.04 ha under vegetables, provision, food or cash crops	10 bearings of trees/ bananas/ plantains in total	1 head of cattle	2 sheep/goats/pigs in total			10 poultry/rabbits in total	1 horse; or 1 donkey		
Saint Lucia	0.05 ha under temporary crops	10 bearing trees; or 100 mats of banana/ plantain in total	1 head of cattle	2 sheep/goats/pigs in total			12 poultry/rabbits in total			

Censuses by Region	Land	Trees	Cattle/buffaloes/camels	Sheep	Goats	Pigs	Poultry	Other livestock	Value/sales	Other
Saint Vincent and the Grenadines	0.04 ha under vegetables, provision, food or cash crops; or 18 sq. m. of greenhouses	10 bearings of trees/bananas/plantains in total	1 head of cattle	2 sheep/goats/pigs in total			12 poultry/rabbits in total	1 horse; or 1 donkey		
Trinidad and Tobago									50% of production sold	
United States									US$1 000 of annual sales	
Virgin Islands. United States									US$500 of annual sales	
AMERICA, SOUTH (5 CENSUSES WITH REPORTED MINIMUM THRESHOLDS)										
Argentina	0.05 ha of total land								50% of production sold	
Chile	0.1 ha under crops; or 0.1 ha of permanent pastures; or 0.1 ha of wood plantations; or 5 ha of wood/forest land		2 heads of large livestock with more than 2 years of age; or 4 heads of large livestock with less than 2 years of age	10 sheep/goats/pigs in total			100 chickens; or 50 geese/ducks/rabbits in total	10 beehives		
Ecuador	0.5 ha of agricultural land									1 product for sale
French Guiana	1 ha of agricultural land; or 0.2 ha under specialized crops;		1 cow							
Uruguay	1 ha of total land									
ASIA (13 CENSUSES WITH REPORTED MINIMUM THRESHOLDS)										
Bangladesh	0.021 ha of total land(4)									
Cyprus	0.1 ha of agricultural land; or 0.05 ha of greenhouses		1 cow; or 2 other large animals (cattle, horses, donkeys, mules)	5 sheep/goats/pigs in total			50	20 beehives		
Iran, Islamic Republic of	0.04 ha under temporary crops; or 0.02 ha under permanent crops; or any area of greenhouses		1 head of cattle/camel in total	2 sheep/goats in total			10	1 beehive; or 1 silkworm		
Jordan	0.1 ha of total land(5)		1 head of cattle; or 3 camels	10	10			5 beehives		
Korea, Rep. of	0.1 ha of land under crops								500 000 Wons(6) of annual sales; or 500 000 Wons of total value of livestock being raised	

Table A8 - Minimum thresholds for inclusion of holdings used in the WCA 2000 round

Censuses by Region	Land	Trees	Cattle/buffaloes/camels	Sheep	Goats	Pigs	Poultry	Other livestock	Value/sales	Other
Lao People's Democratic Republic	0.02 ha of agricultural land;		2 cattle/buffaloes in total		5 goats/pigs in total		20			
Lebanon	0.05 ha of non-irrigated arable land; or 0.025 ha of irrigated arable land; or 100 sq. m. of greenhouses		1 head of cattle		7 goats/pigs in total		15			
Myanmar	0.041 ha under crops(7)		2 heads of large livestock	4 sheep/goats/pigs in total			30 chickens/ducks in total			
Nepal	0.012 ha under crops in hill / mountain districts and 0.014 ha under crops in lowland districts		2 cattle/buffaloes in total	5 sheep/goats in total			20			
Pakistan			1 head of cattle	5 sheep/goats in total						
Philippines	0.1 ha under crops		20 heads of livestock in total				20			
Qatar	Any area of land used for agriculture		1 cow; or 1 camel	5 sheep/goats in total			50			
Sri Lanka	0.11 ha(8) of total land								50% of production sold	
EUROPE (20 CENSUSES WITH REPORTED MINIMUM THRESHOLDS)										
Austria	1 ha of agricultural land; or 0.25 ha of commercial vineyards; or 0.15 ha or orchards; or 0.1 ha under berries/vegetables; or 0.1 ha of greenhouses; or any area under mushrooms		3 heads of cattle	10 sheep/goats in total		5	100			
Czech Republic	1 ha of agricultural land; or 0.15 ha under permanent crops; or 0.05 ha of greenhouses		1 head of cattle	4 sheep/goats in total		2	50	100 fur-bearing animals		
Denmark	1 ha of agricultural land								Standard Gross Margin of €4 000 at 1990 prices	
Estonia	1 ha of agricultural or forest land								50% of production sold	0.3 ha of fish ponds
Finland	1 ha of arable land								Standard Gross Margin of 1 European Size Unit	
France	1 ha of agricultural land; or 0.2 ha under specialized crops									A certain number of livestock; or a certain level of agricultural output

Censuses by Region	Land	Trees	Cattle/buffaloes/camels	Sheep	Goats	Pigs	Poultry	Other livestock	Value/ sales	Other
Germany	2 ha of agricultural land; or 10 ha of wooded area; or 0.3ha of special crops(9)		8 bovines	20		8	200 laying hens/ broilers/geese in total			
Greece	0.1 ha of agricultural land; or 0.05 ha of greenhouses		1 cow; or 2 other large livestock	5 sheep/goats/pigs in total			50	20 beehives		
Hungary	0.15 ha of total land; or intensive horticultural activities under glass		1 head of cattle	1	1	1	50 poultry; or 1 ostrich	1 donkey; or 1 mule; or a stock of 25 rabbits, fur-bearing animals or pigeons for meat		
Ireland	1 ha of agricultural land									Intensive agricultural production
Latvia	1 ha of agricultural land								1 000 LVL(10) of annual sales	
Lithuania	1 ha of agricultural land								40 MSL (Minimum Standards of Living) of income	
Luxembourg	1 ha of cultivated land; or 1 ha of vineyards									
Netherlands									Standard Gross Margin of 3 European Size Units	
Norway	0.5 ha of agricultural land; or 300 sq. m. of greenhouses; or 0.2 ha under field grown vegetables; or 0.1 ha of field grown berries; or 0.1 ha under fruit trees; or 0.1 ha of nurseries		10 heads of cattle	25 over 1 year of age	10 over 1 year of age	5 breeding pigs and/ or 200 other pigs	1 000 laying hens and/or chickens reared for egg production; or 5 000 chickens for meat production			
Poland	0.1 ha of agricultural land		1 head of cattle	3 sheep/goats in total		1 sow; or 5 heads of pigs	30	1 horse; or 5 females of fur animals		
Portugal	1 ha of agricultural land; or 0.5 ha under potatoes or other extensively cultivated crops; or 0.2 ha of industrial crops or orchards; or 0.1 ha of specialized cultures; or 0.05 ha of flowers or greenhouses or nurseries		1 cow; or 1 breeding bull; or 2 heads of cattle two years old	6 sheep/goats in total		1 swine; or 3 pigs for fattening	100	10 beehives; or 10 female rabbits		

▶ Table A8 - Minimum thresholds for inclusion of holdings used in the WCA 2000 round

Censuses by Region	Land	Trees	Cattle/buffaloes/camels	Sheep	Goats	Pigs	Poultry	Other livestock	Value/ sales	Other
Serbia	0.1 ha of cultivable land		1 cow and 1 calf; or 1 cow and 1 heifer; 1 cow and 2 fully grown heads of small livestock	5 fully grown		3 fully grown	50 fully grown	20 beehives		4 fully grown heads of sheep and pigs in total
Slovenia	1 ha of agricultural land; or 0.1 ha of agricultural land and 0.9 ha of wood/forest; or 0.5 ha of arable land/ kitchen garden in total; or 0.3 ha under permanent crops; or 0.1 ha of orchards plantation		2 heads of cattle					50 beehives		Market producer of vegetables
Spain	0.1 ha of total land		1 head of cattle	6 sheep/goats in total			50	10 beehives; or 10 female rabbits; or 2 horses/asses/ mules/pigs in total		
Sweden	2.1 ha of arable land; or 200 sq. m. of greenhouse; or 0.25 ha of outdoor cultivation		50 cows or 250 heads of cattle	50 ewes		50 sows or 250 pigs	1000			
Oceania (5 censuses with reported minimum thresholds)										
Australia									5 000 AUD(11) of Estimated value of agricultural operations	
Cook Islands	0.05 ha under garden crops	10 crop trees								
Guam									US$1 000 of annual sales	
New Caledonia										350 points(12)
Northern Mariana Islands									US$1 000 of annual sales	

1 The actual threshold was 12 sahms
2 The actual threshold was 1 acre
3 The actual threshold was 4 400 square feet
4 The actual threshold was 0.05 acres
5 The actual threshold was 1 dunum
6 500 000 Wons = about US$427 at the time of the census
7 The actual threshold was 0.1 acre
8 The actual threshold was 40 perches

9 Vineyards/hops/tobacco/tree nurseries/outdoor flowers/market gardening and cultivation under cover/medicinal plants
10 1 LVL (Latvian Lat) = US$1.58 at the time of the census
11 1 AUD (Australian Dollar)= US$0.6 at the time of the census
12 Points calculated as follows: 100 points for 1 ha under pasture-land, 100 points for 1 milk cow, 100 points for 1 sow, 20 points for 100 sq. m. under vegetables, 20 points for 1 bee-hive, 2 points for 1 laying-hen, etc.

Table A9 - Coverage of census item categories in the WCA 2000 round

Countries by region	01 Identification	02 General characteristics	03 Demographic characteristics	04 Employment	05 Land and Water	06 Crops	07 Livestock	08 Machinery and equipment	09 Buildings and other structures	10 Other activities
AFRICA (25 REPORTED CENSUSES)										
Algeria	✓	✓	✓	✓	✓	✓	✓	✓		
Botswana	✓	✓	✓	✓	✓	✓	✓	✓		
Cape Verde	✓	✓	✓	✓	✓	✓	✓	✓		
Comoros	✓	✓	✓	✓	✓	✓	✓	✓	✓	
Côte d'Ivoire	✓		✓	✓	✓	✓	✓	✓	✓	
Egypt	✓	✓	✓	✓	✓	✓	✓	✓	✓	✓
Ethiopia	✓		✓		✓	✓	✓		✓	
Gambia	✓		✓			✓	✓			
Guinea	✓	✓	✓	✓	✓	✓	✓	✓		
Lesotho	✓	✓	✓	✓		✓	✓			
Libyan Arab Jamahiriya	✓			✓		✓	✓			
Madagascar	✓	✓	✓	✓	✓	✓	✓	✓	✓	✓
Mali	✓	✓	✓	✓	✓	✓	✓	✓	✓	
Morocco	✓	✓	✓	✓	✓	✓	✓	✓		
Mozambique	✓	✓	✓	✓	✓	✓	✓	✓	✓	
Namibia	✓		✓	✓	✓	✓	✓			
Réunion	✓	✓	✓	✓	✓	✓	✓	✓	✓	✓
Senegal	✓		✓	✓	✓	✓	✓	✓		
Seychelles	✓		✓	✓	✓		✓			
South Africa	✓	✓	✓	✓	✓	✓	✓	✓	✓	✓
Tanzania, United Republic of	✓	✓	✓	✓	✓	✓	✓	✓		✓
Togo	✓		✓	✓	✓	✓	✓	✓		
Tunisia	✓	✓	✓	✓	✓	✓	✓	✓	✓	✓
Uganda	✓		✓	✓	✓	✓	✓			✓
Zambia	✓		✓	✓		✓	✓			
Subtotal Africa	**25**	**15**	**24**	**23**	**21**	**24**	**25**	**17**	**10**	**7**
AMERICA, NORTH & CENTRAL (14 REPORTED CENSUSES)										
Canada	✓	✓	✓	✓	✓	✓	✓	✓		
Guadeloupe	✓	✓	✓	✓	✓	✓	✓	✓	✓	✓
Guatemala	✓	✓	✓	✓	✓	✓	✓	✓	✓	✓
Jamaica	✓				✓	✓	✓		✓	
Martinique	✓	✓	✓	✓	✓	✓	✓	✓	✓	✓
Nicaragua	✓	✓	✓	✓	✓		✓		✓	
Panama	✓		✓		✓	✓			✓	
Puerto Rico	✓	✓	✓	✓	✓	✓	✓	✓	✓	✓
Saint Kitts and Nevis	✓	✓	✓		✓	✓	✓		✓	✓
Saint Lucia	✓	✓	✓	✓	✓	✓	✓	✓		
Saint Vincent	✓	✓	✓	✓	✓	✓	✓		✓	
Trinidad and Tobago	✓	✓	✓	✓	✓	✓	✓	✓	✓	✓
United States of America	✓	✓	✓	✓	✓	✓	✓	✓	✓	✓
Virgin Islands, United States	✓	✓	✓	✓	✓	✓	✓	✓		✓
Subtotal America North & Central	**14**	**12**	**13**	**11**	**14**	**13**	**13**	**9**	**11**	**8**

Table A9 - Coverage of census item categories in the WCA 2000 round

Countries by region	Census Item Category									
	01 Identification	02 General characteristics	03 Demographic characteristics	04 Employment	05 Land and Water	06 Crops	07 Livestock	08 Machinery and equipment	09 Buildings and other structures	10 Other activities
AMERICA, SOUTH (8 REPORTED CENSUSES)										
Argentina	✓	✓		✓	✓	✓	✓	✓		✓
Brazil	✓	✓	✓	✓	✓	✓	✓		✓	
Chile	✓	✓	✓	✓	✓	✓	✓	✓	✓	✓
Colombia	✓	✓			✓	✓	✓			✓
Ecuador	✓	✓	✓	✓	✓	✓	✓	✓	✓	✓
French Guiana	✓	✓	✓	✓	✓	✓	✓	✓	✓	✓
Uruguay	✓	✓	✓	✓	✓	✓	✓	✓	✓	✓
Venezuela, Bolivarian Republic of	✓	✓	✓	✓	✓	✓	✓		✓	
Subtotal America, South	8	8	6	7	8	8	8	5	6	6
ASIA (29 REPORTED CENSUSES)										
Afghanistan	✓						✓			
Azerbaijan	✓	✓	✓	✓	✓	✓	✓	✓	✓	✓
Bangladesh	✓		✓	✓	✓	✓	✓	✓		
Bhutan	✓					✓	✓			
China	✓	✓	✓	✓	✓	✓	✓	✓	✓	✓
Cyprus	✓	✓	✓	✓	✓	✓	✓	✓	✓	✓
Georgia	✓	✓	✓		✓	✓	✓	✓		
India	✓	✓	✓		✓	✓	✓		✓	
Indonesia	✓				✓	✓	✓		✓	
Iran, Islamic Republic of	✓	✓			✓	✓	✓			
Japan	✓	✓	✓	✓	✓	✓	✓	✓		✓
Jordan	✓	✓	✓	✓	✓	✓	✓			
Korea, Republic of	✓		✓							
Kyrgyzstan	✓	✓	✓	✓	✓	✓	✓	✓	✓	
Lao People's Democratic Republic	✓	✓	✓	✓	✓	✓	✓	✓		✓
Lebanon	✓	✓	✓	✓	✓	✓	✓	✓		
Malaysia	✓		✓		✓	✓				✓
Mongolia	✓						✓			
Myanmar	✓	✓	✓	✓	✓	✓	✓	✓		✓
Nepal	✓	✓	✓	✓	✓	✓	✓	✓	✓	✓
Pakistan	✓	✓	✓	✓	✓	✓	✓		✓	
Philippine	✓	✓	✓	✓	✓	✓	✓	✓		✓
Qatar	✓		✓	✓	✓	✓	✓		✓	
Saudi Arabia	✓	✓	✓	✓	✓	✓	✓		✓	
Sri Lanka	✓	✓	✓		✓	✓	✓	✓		
Thailand	✓	✓	✓	✓	✓	✓	✓	✓	✓	✓
Turkey	✓	✓		✓	✓	✓	✓	✓	✓	
Viet Nam	✓		✓	✓	✓					
Yemen	✓	✓		✓	✓	✓	✓	✓		
Subtotal Asia	29	21	22	19	25	25	26	16	13	12

Table A9 - Coverage of census item categories in the WCA 2000 round

Countries by region	Census Item Category									
	01 Identification	02 General characteristics	03 Demographic characteristics	04 Employment	05 Land and Water	06 Crops	07 Livestock	08 Machinery and equipment	09 Buildings and other structures	10 Other activities
EUROPE (29 REPORTED CENSUSES)										
Albania	✓	✓	✓	✓	✓	✓	✓	✓	✓	
Austria	✓	✓	✓	✓	✓	✓	✓	✓	✓	
Belgium	✓	✓	✓	✓	✓	✓	✓	✓	✓	✓
Croatia	✓	✓	✓	✓	✓	✓	✓	✓	✓	
Czech Republic	✓	✓			✓	✓	✓		✓	
Denmark	✓	✓	✓	✓	✓	✓	✓	✓	✓	
Estonia	✓	✓	✓	✓	✓	✓	✓		✓	✓
Finland	✓	✓	✓	✓	✓	✓	✓	✓	✓	
France	✓	✓	✓	✓	✓	✓	✓	✓	✓	✓
Germany	✓	✓	✓	✓	✓	✓	✓		✓	
Greece	✓	✓	✓	✓	✓	✓	✓	✓	✓	
Hungary	✓	✓	✓	✓	✓	✓	✓	✓	✓	✓
Ireland	✓	✓	✓	✓	✓	✓	✓	✓	✓	
Italy	✓	✓	✓	✓	✓	✓	✓	✓	✓	✓
Latvia	✓	✓			✓	✓	✓		✓	
Lithuania	✓	✓	✓	✓	✓	✓	✓	✓	✓	✓
Luxembourg	✓	✓	✓	✓	✓	✓	✓	✓	✓	
Malta	✓	✓	✓	✓	✓	✓	✓	✓	✓	
Netherlands	✓	✓	✓	✓	✓	✓	✓	✓	✓	
Norway	✓	✓	✓	✓	✓	✓	✓	✓	✓	
Poland	✓	✓	✓	✓	✓	✓	✓	✓	✓	
Portugal	✓	✓	✓	✓	✓	✓	✓	✓	✓	✓
Romania	✓	✓	✓	✓	✓	✓	✓	✓	✓	✓
Serbia	✓	✓	✓	✓	✓		✓			
Slovakia	✓	✓	✓		✓	✓	✓	✓	✓	
Slovenia	✓	✓	✓	✓	✓	✓	✓	✓	✓	✓
Spain	✓	✓	✓	✓	✓	✓	✓	✓	✓	
Sweden	✓	✓	✓	✓	✓	✓	✓	✓	✓	
United Kingdom	✓	✓	✓	✓	✓	✓	✓		✓	✓
Subtotal Europe	29	29	27	27	29	28	29	23	28	10
OCEANIA (9 REPORTED CENSUSES)										
American Samoa	✓	✓	✓	✓	✓	✓	✓	✓		✓
Australia	✓				✓	✓	✓		✓	
Cooks Islands	✓		✓	✓	✓	✓	✓		✓	✓
Guam	✓	✓	✓	✓	✓	✓	✓	✓		✓
New Caledonia	✓	✓	✓	✓	✓	✓	✓		✓	✓
New Zealand	✓				✓	✓	✓		✓	
Northern Mariana Islands	✓	✓	✓	✓	✓	✓	✓	✓		✓
Samoa	✓	✓	✓	✓	✓	✓	✓	✓	✓	✓
Tonga	✓	✓	✓	✓	✓	✓	✓	✓	✓	✓
Sub total Oceania	9	6	7	7	9	9	9	6	6	7
TOTAL	114	91	99	94	106	107	110	76	74	50

REFERENCES

FAO. 1995. *Programme for the World Census of Agriculture 2000*. FAO Statistical Development Series 5. Rome.

FAO. 1996a. *Conducting agricultural censuses and surveys*, FAO Statistical Development Series 6. Rome.

FAO. 1996b. *Multiple frame agricultural surveys*. Vol. 1. *Current surveys based on area and list sampling methods*. FAO Statistical Development Series 7. Rome.

FAO. 1997a. *Guidelines on employment. Supplement for the Programme for the World Census of Agriculture 2000*. FAO Statistical Development Series 5a. Rome.

FAO. 1997b. *Guidelines on the collection of structural aquaculture statistics. Supplement for the Programme for the World Census of Agriculture 2000*. FAO Statistical Development Series 5b. Rome.

FAO. 2005. *A system of integrated agricultural censuses and surveys*. Vol. 1. *World Programme for the Census of Agriculture 2010*. FAO Statistical Development Series 11. Rome.

FAO. 2010. *2000 World Census of Agriculture. Main Results and Metadata by Country (1996-2005)*. FAO Statistical Development Series 12. Rome.

FAO. 2013. *2000 World Census of Agriculture. Analysis and International Comparisons of Results (1996 – 2005)*. FAO Statistical Development Series 13. Rome.